Terminal Town:

An Illustrated Guide to Chicago's Airports, Bus Depots, Train Stations, and Steamship Landings

1939 - Present

An Illustrated Guide to Chicago's Airports, Bus Depots, Train Stations, and Steamship Landings 1939 - Present

By Joseph P. Schwieterman

Joseph P. Schwieterman PhD

Chicago, Illinois

LAKE FOREST
COLLEGE

LAKE
FOREST
COLLEGE
PRESS

Cover photos:

Top: O'Hare International Airport, 1963 (Chicago History Museum, Photographer: Bill Engdahl, HB-25500-B2)

Bottom: (from left): LaSalle Street Station, 1974 (Craig Bluschke photo); Meigs Field, 2003 (Lawrence Okrent photo); Grand Central Station, 1966 (George Drury photo, courtesy of Kevin Holland); Chicago Trailways Station, 1986 (Mel Bernero collection)

Lake Forest College
555 N. Sheridan Road
Lake Forest, IL 60045

lakeforest.edu/lfcpress

Copyright © Lake Forest College Press 2014

Lake Forest College Press publishes in the broad spaces of Chicago studies. Our imprint, &NOW Books, publishes innovative and conceptual literature, and serves as the publishing arm of the &NOW writers' conference and organization.

ISBN: 978-0-9823156-9-9

Book design by Emma C. O'Hagan

Printed in the United States

LAKE FOREST
COLLEGE

LAKE
FOREST
COLLEGE
PRESS

ACKNOWLEDGEMENTS
Terminal Town

This volume would not have been possible without the editorial and research support of Kevin J. Holland; Norman Carlson, Dave Hoffman, and Bruce Moffatt of the Shore Line Interurban Historical Society; William Molony of the Blackhawk Chapter of the National Railway Historical Society; and members of the Midway Airport Historians, including Pat Bukiri, David Kent, Christopher Lynch, and Bob Russo. I would also like to recognize Fred Ash, Barry Chrenen, Rick Harnish, Graham Garfield, Art Peterson, F.K. Plous, Jeffrey Sriver, William Vandervoort, I.E. Quastler, and John Zukowski for their guidance and insights.

Many individuals at DePaul University also lent a hand, including Xhoana Ahmeti, Susan Aaron, Lauren Fischer, Justin Kohls, John Hedrick, Laurie Marston, Paige Largent, Dana Nelson, Sara Lepro, Mollie Pelon, Ryan Forst, Marisa Schulz, and Kate Witherspoon.

The custom maps appearing in this volume were created by Dennis McClendon of Chicago Cartographics and Rick Johnson of Kalmbach Publishing Co. The author thanks these talented cartographers and the staff of *Trains* magazine for permission to use the Kalmbach publication's base map of U.S. rail-passenger routes. Readers interested in events and multimedia productions accompanying this book, or links to photographs used from Creative Commons, should visit www.terminaltown.org.

Finally, I also recognize the staff at Lake Forest College Press, including Emma O'Hagan, Leslie Taylor, Nicole FioRito, and Davis Schneiderman, director of Lake Forest College Press.

TABLE OF CONTENTS
Terminal Town

INTRODUCTION
Terminal Town

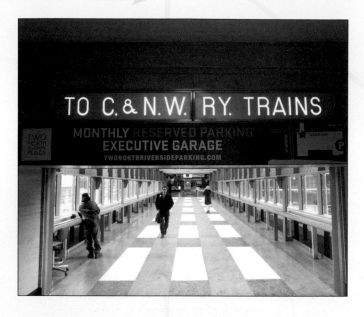

Chicago's system of passenger transportation terminals, famously complex and constantly changing, has for more than a century been a defining feature of its cosmopolitan character. Whether serving those traveling by aircraft, boat, bus, or train, these places have given the "Windy City" its reputation as the epicenter of America's passenger transportation network.

Yet the full extent of this system has gone unappreciated and misunderstood by many observers. Some of the most remarkable terminals are ignored in publications about the city's history and largely unknown to contemporary travelers and transportation professionals. To provide readers a grasp of the system's extraordinary proportions, this book highlights 48 places in the metropolitan region that have been termini or important connecting points for intercity passengers since 1939. The various chapters take readers to the nerve centers of a transportation system that shoulders an enormous burden: moving people across the continent.

Chicago deserves to be the focus of such a book, if for nothing else than its top ranking in key areas of passenger travel. From the beginning of the twentieth century through 1969, Chicago had *six* major downtown railroad stations—twice as many as any other large American city. Between the 1930s and 1998, Chicago was home to the world's busiest airport (initially Midway Airport and later O'Hare International Airport). Passengers boarding in Chicago could—and can still—fly directly to more major American cities than from any other city in the country. From 1953 to 1989, Chicago was home to the nation's largest independently operated bus depot—the Chicago Greyhound Station—a facility that offered direct bus service to more places than from any other American city.

Evaluating this ponderous system of terminals reveals both the flattering and less-than-flattering sides of Chicago's transportation heritage. For generations, travelers who simply want to go from Point A to Point B have looked at Chicago's terminal system with both a sense of awe and a sense of dread. The city gives them innumerable travel alternatives while also requiring them to contend with some of the country's most notorious bottlenecks. When the city's population was near its peak in 1955, buses, trains, planes, and ships arriving in the metropolitan area terminated at a bewildering 20 different locations.[1] Another five transfer points, some on the region's periphery, offered opportunities for timesaving connections for those passing through the region.

Readers may immediately be drawn to the chapters featuring the great icons of American transportation—Chicago Union Station, Midway, and O'Hare, to name only a few. Smaller and less-publicized locations, however, also deserve attention. Englewood Union Station on the South Side once had direct service to more of America's 100 largest cities than was available from any other train depot in the United States. Sky Harbor Airport in Northbrook, was home to the world's first "scheduled air taxi shuttle." Tiny Meigs Field on the Chicago waterfront boasted six carriers operating 122 daily passenger flights—the most ever at any downtown airport in the country.

The chapters featuring each of the stations and terminals are intentionally brief and focus primarily on the *transportation* role of each facility, to allow readers to appreciate the gestalt of Chicago passenger transportation. The first part of the book describes the downtown bus, train, and steamship terminals, and is followed by a review of the region's outlying termini and time-saving connecting points. Major airports and the smaller airfields used for short-hop air taxi flights are showcased in the latter part of the book, followed by an assessment of profound changes looming on the horizon.

Readers seeking detail on architecture or design, or more about the particular transportation services available, may wish to consult the bibliography at the end of this volume. Those interested in the many calculations presented in the pages that follow should consult the author's companion paper summarized in Appendix III. By design, the volume excludes from consideration the region's many local stations that were neither endpoints for intercity bus or train routes nor important transfer points for passengers traveling through the region.

As this book illustrates, the place so aptly described as the Windy City, the City of the Big Shoulders, and, as rhapsodized by Frank Sinatra, "that toddlin' town," is equally deserving of being called Terminal Town. ▪

(Opposite) An illuminated sign evokes the memory of the Chicago & North Western Railway at the entrance to the footbridge over Canal St. leading to Ogilvie Station. (Author's photo)

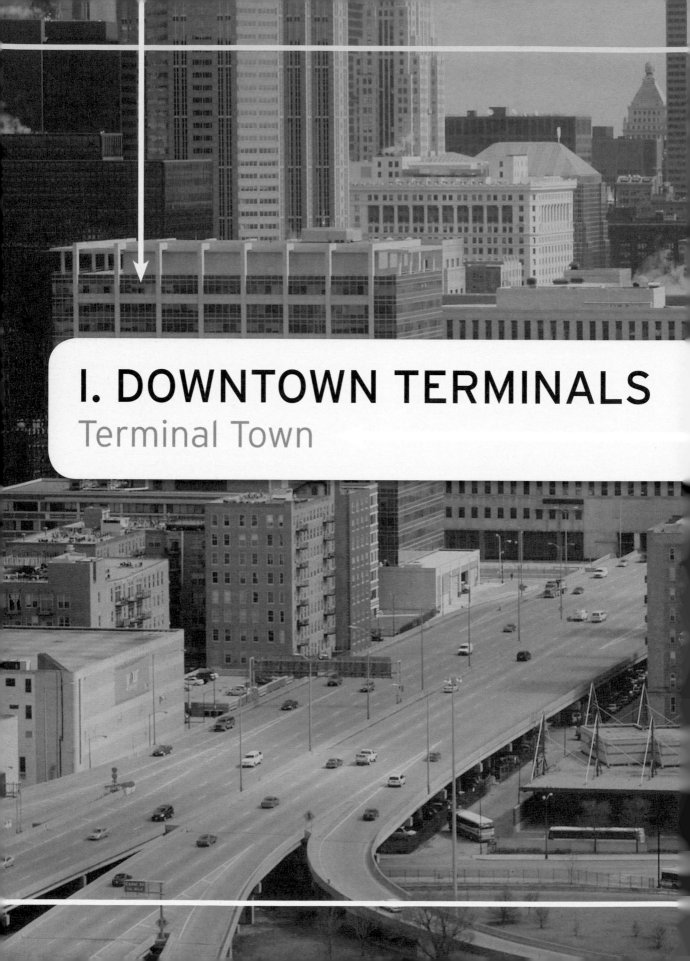

I. DOWNTOWN TERMINALS
Terminal Town

DOWNTOWN RAILROAD TERMINALS
Terminal Town

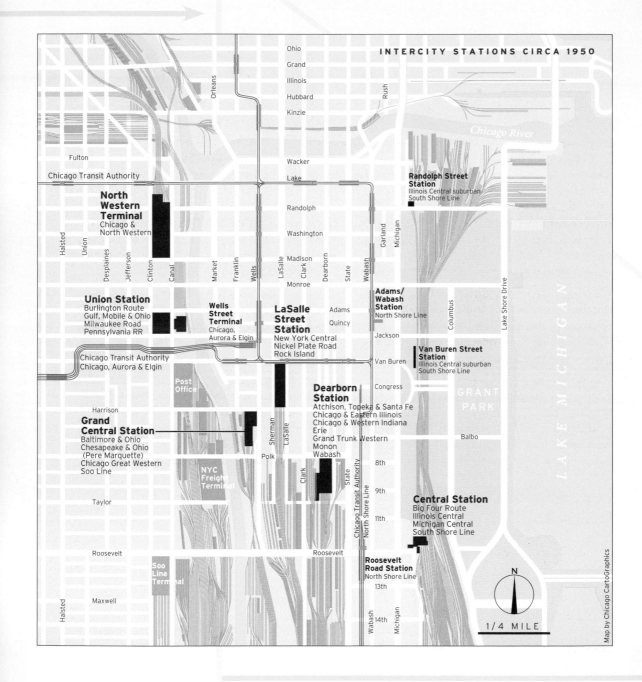

INTERCITY STATIONS CIRCA 1950

Randolph Street Station
Illinois Central suburban
South Shore Line

North Western Terminal
Chicago & North Western

Union Station
Burlington Route
Gulf, Mobile & Ohio
Milwaukee Road
Pennsylvania RR

Chicago Transit Authority
Chicago, Aurora & Elgin

Wells Street Terminal
Chicago, Aurora & Elgin

LaSalle Street Station
New York Central
Nickel Plate Road
Rock Island

Adams/Wabash Station
North Shore Line

Van Buren Street Station
Illinois Central suburban
South Shore Line

Grand Central Station
Baltimore & Ohio
Chesapeake & Ohio
(Pere Marquette)
Chicago Great Western
Soo Line

Dearborn Station
Atchison, Topeka & Santa Fe
Chicago & Eastern Illinois
Chicago & Western Indiana
Erie
Grand Trunk Western
Monon
Wabash

Central Station
Big Four Route
Illinois Central
Michigan Central
South Shore Line

Roosevelt Road Station
North Shore Line

Post Office

NYC Freight Terminal

Soo Line Terminal

Chicago Transit Authority

North Shore Line

GRANT PARK

LAKE MICHIGAN

Chicago River

N

1/4 MILE

Map by Chicago CartoGraphics

Chicago's downtown terminals were for generations the showpieces of its vast transportation inventory. From the beginning of the twentieth century through the tumultuous era of passenger-train elimination in the late 1960s, the city had six large stations. Central Station, Dearborn Station, Grand Central Station, LaSalle Street Station, North Western Terminal, and Union Station—the "Big Six"—were the endpoints of almost all passengers arriving into the metropolitan region for more than 40 years.

Real estate leader Charles Nichols described the stations as "the mouths that fed" the city in a speech to commercial leaders in the 1920s.[2] His metaphor was apt: In 1928, more than 440 intercity trains and hundreds of commuter trains rumbled in and out of the city's six downtown terminals. The number of trains diminished somewhat during the Great Depression but still stood at 356 in 1942, when traffic surged in response to the country's deepening involvement in World War II.[3] Seventeen railroads operated trains into Chicago's downtown terminals as the country mobilized for war, carrying travelers directly to 82 of the 100 largest cities in the United States.[4] The largest city that could not be reached directly from Chicago by train, San Antonio, ranked only 25th in population at the time.

Passengers departing downtown Chicago had a choice of railroads to most major cities. Five different carriers (in some cases with connecting railroads) operated between Chicago and Denver, Kansas City, and the Twin Cities, as well as between Chicago and metropolitan New York in 1945. Four operated directly to Omaha and St. Louis. Three operated to Detroit, Jacksonville, Indianapolis, Los Angeles, Miami, and the San Francisco Bay Area.

A cosmopolitan mix of services allowed travelers to arrange itineraries based on the terminal from which they wanted to depart, much like flyers today choose between Midway and O'Hare. Those heading to Denver and Detroit could depart from four different downtown stations, while those destined for metropolitan New York could depart from five.[5] Those bound for Kansas City, Indianapolis, Los Angeles, Miami, and St. Louis could choose from three. Rarely were departure options from Chicago to a major destination limited to a single terminal.

THE BURDEN OF MAKING TRANSFERS

The difficulty of navigating such an elaborate terminal system brought much notoriety to the city. Making transfers between the terminals was the source of particular agitation. Exhorting the industry to simplify travel arrangements, Robert Young, chairman of the Chesapeake & Ohio Railway, famously published an advertisement in newspapers and magazines in 1947 with the slogan "A hog can cross the country without changing trains—but you can't."[6] Young's comment resonated with travelers weary of the hassles of transferring between trains and stations in Chicago.

(Opposite) This map shows Chicago's six major downtown stations as of 1945—Central, Dearborn, Grand Central, LaSalle Steet, Union, and North Western (labeled in red)—and the numerous other stations used by the city's commuter railroads, the Chicago Transit Authority's "L" system, and electric interurban railways. A multiplicity of routes, yards, and spurs supported this network of stations. The stations labeled in red were endpoints of regularly scheduled long-distance passenger trains while those in black were strictly used by commuter or suburban providers, or interurban railways, after 1939.

Transfers between terminals were especially cumbersome for the elderly, families with children, and those toting heavy luggage. Walking between stations, while always an option, was often not as easy as one might expect, as the city's terminals were separated by as much as a mile and a half.[7] Nor was transferring between stations usually as simple as using rapid transit. Only Union and LaSalle were next to the "L," and the former's nearby stop closed in 1958. North Western Terminal was equipped with a walkway to the nearby Clinton L stop in 1970. This cleverly named link—the "Northwest Passage"—came too late to be of much value to intercity travelers, was on the opposite end of the station from the central waiting room, and lasted only until 1990.

Not every traveler passing through the city, of course, had to transfer between terminals. Dearborn, LaSalle Street, and Union stations were relatively efficient hubs at which trains from the East met those to the West, making same-station connections pervasive. But, the decentralized arrangement of the city's terminals made transfers between stations a necessity for most people.

Efforts to consolidate terminals were repeatedly stymied by issues of complexity and cost. Daniel H. Burnham and Edward H. Bennett understood that the arrangement of stations was deeply problematic, but were quite ambiguous in their famed *Plan of Chicago* about how to deal with this issue. Some drawings in the lavishly illustrated 1909 plan show a handful of consolidated terminals, while others show as many as seven spread throughout the downtown region. Not only were the grand visions of this era never realized; even more modest efforts, such as proposals to close Dearborn and divide its trains among the other stations, never came to be.

Several milestones were nonetheless achieved following the completion of Burnham and Bennett's plan. Chicago & North Western replaced its old Wells Street Station (on the site of the present Merchandise Mart) with the magnificent North Western Terminal on Madison St. in 1911. In 1925, a new Union Station, designed by Graham, Anderson, Probst & White, opened, replacing the older depot of the same name. This Beaux Arts landmark was the largest terminal ever built in Chicago, and was exceeded in size in this country only by its counterpart in Kansas City, and New York's Grand Central and Pennsylvania stations.

Some expected—optimistically, as it turns out—that Union Station would absorb a great deal more traffic from the city's older stations. In 1953, the Chicago Plan Commission proposed amalgamating Dearborn, Grand Central, and LaSalle Street stations into one giant station south of Harrison St. (see drawing on next page). Others later placed hope in relocating operations at these stations to a new 20-track depot on the site of Central Station.[8] But these efforts fared no better than past plans, and the consolidation movement was left dead in its tracks. Like it or not, Chicago remained a six-station town until most of the passenger trains were eliminated.

Passengers, railroads, and profit-seeking businesspeople did their best to minimize the associated inconveniences. Parmelee Transfer Co. shuttled passengers and their trunks and luggage between terminals, initially in horse-drawn carts but later in vans and limousines. Taxicabs and city buses were also popular options. On a few routes, railroad companies transferred sleeping cars between trains departing from different stations while allowing passengers to remain on board. Travelers found this far from ideal, however, since they were confined on board their sleeping car during a process that could take hours.

Some passengers studiously avoided the downtown terminals by making connections at junctions outside of the central business district. (These are discussed in the chapter "Timesaving Connecting Points"). At no time during the era of privately operated passenger trains, however, did entire scheduled trains actually operate *through* Chicago. Early in its existence, Amtrak broke from tradition by operating a train through the city. Starting in 1971, trains operated through Union Station between Milwaukee and St. Louis, using one of only two run-through tracks within the station. This operation lasted a little more than two years.

The railroads rolled into the post-World War II era with guarded optimism about the passenger business. New streamlined passenger equipment improved the image and speed of travel. Air conditioning, reserved seating, "Slumbercoach" cars (giving budget-minded travelers the convenience of private rooms), and dome cars that allowed for better sightseeing appeared to bode well for the future. A great exhibition, the Chicago Railroad Fair, was held along the south lakefront during 1948 and 1949 to commemorate a century of railroading in

(Opposite) Trains of the Rock Island Lines pause under LaSalle Street Station's train shed on March 5, 1975. At the time, the railroad still operated the *Peoria Rocket* and *Quad Cities Rocket*—the last survivors of the once-vast system of privately operated intercity passenger trains serving Chicago. The *Rockets* remained in service until December 31, 1978, on account of the Rock Island's decision to remain independent of Amtrak. (Verne Brummel photo)

DOWNTOWN RAILROAD TERMINALS
Terminal Town

the city. As late as 1956—on the eve of the Interstate highway system and the introduction of Boeing 707 airliners—rail passengers beginning their trips in Chicago could still travel directly to more than three-quarters of the top-100 cities that had been available to them in 1942.[9]

The failed efforts to consolidate terminals, however, weighed heavily on the railroads. As the financial performance of their passenger trains worsened, these massive edifices were increasingly seen as a financial drain. The geographic and historical factors that had saddled the railroads with less-than-optimal terminal situations grew more problematic as competition from other modes intensified, labor costs escalated, and property taxes crept higher. All six of the great stations began to suffer from neglect.

The inability of railroads to bring passengers to the heart of downtown certainly didn't help matters. Chicago's great stations were less conveniently located than those of New York City, which were surrounded by prime commercial real estate and high-density development from their earliest days.[10] Having stations on the periphery of downtown Chicago did not pose serious problems through the late 1920s, when most commercial activity was south of Madison St. Gradually, however, commerce rapidly spread north, streetcar service fell out of favor, and competition from motor vehicles escalated, making the locations of the stations a growing inconvenience.

By the early 1960s, Central, Dearborn, and Grand Central seemed isolated on downtown's southern fringe. "[T]he two stations with 'central' in their names actually are on the rim of the downtown section," complained the *Chicago Tribune*'s William R. Miner, who also lightheartedly observed that the names of the stations were misleading.[11] "And the New York Central, which the unenlightened might expect to find in a station labeled 'central,' actually comes into the LaSalle Street station with its main trains." No doubt, many travelers shared Miner's frustrations.

FROM SIX TO ONE

As the red ink from operating passenger trains increased, aggravated by a sharp drop in express and mail shipments, the pressure to consolidate stations became overwhelming. Between 1969 and 1978, Chicago went from having six terminals for conventional long-distance trains to just one.

Grand Central closed in 1969. More than two-thirds of the city's long-distance trains were eliminated upon the start up of Amtrak on May 1, 1971, ending intercity service at Dearborn Station and North Western Terminal. The government-subsidized carrier made Union Station its primary Midwestern hub. Direct service to hundreds of communities was eliminated, overnight, with Amtrak's creation, but passengers were afforded better connecting opportunities by virtue of having a consolidated terminal. During the difficult transition period that ensued, Central and LaSalle bid farewell to their last intercity trains in 1972 and 1978, respectively.

However, the near-absence of run-through tracks became a more persistent problem at Union Station. Most arriving trains serving Union Station were pulled out for servicing by a switch locomotive, consuming time and track space. Any hope of a large-scale reconfiguration of the track to allow for extensive run-through operations disappeared when the concourse was razed in 1969 to build an office complex above.

Today, Central and Grand Central are gone, although visitors can still find evidence of their profound physical presence. LaSalle Street Station and North Western Terminal, while dramatically altered, continue to serve commuter trains linking the city and nearby suburbs. Dearborn's historic head house survives in adaptive reuse but remains devoid of a transportation role. Yet the hope of improving Union Station is growing in the wake of an ambitous master plan. The legacy of these magnificent stations is presented alphabetically on the following pages. ■

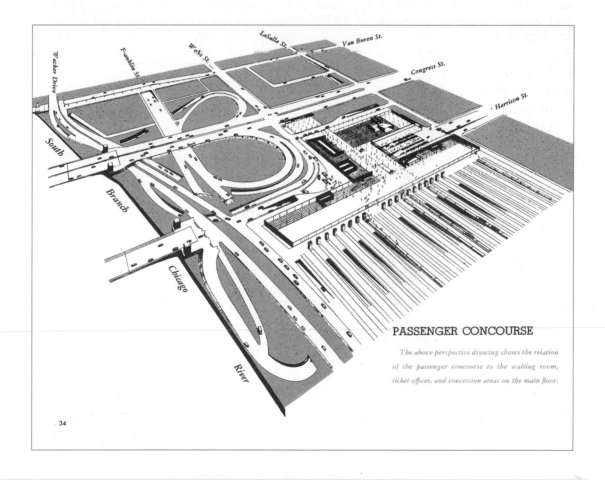

PASSENGER CONCOURSE

The above perspective drawing shows the relation of the passenger concourse to the waiting room, ticket offices, and concession areas on the main floor.

34

This drawing shows the layout of a 1953 Chicago Plan Commission proposal to consolidate Dearborn, Grand Central, and LaSalle Street stations into a giant stub-end depot bounded by LaSalle St. and an extension of Wacker Dr. Motorists would have convenient airport-style drop-off on curved roadway created by modifying Harrison St. and other roads. (Chicago Plan Commission, courtesy of Chicago Cartographics)

CENTRAL STATION
Michigan Ave. and 11th Pl.

Illinois Central's *Illini*, an afternoon departure bound for Carbondale, IL., pokes out from Central Station's modernized train shed on April 9, 1971, less than a month before Amtrak assumed responsibility for the nation's intercity passenger trains. The IC station's massive head house, clock tower, and rooftop billboard can be seen in the upper-left portion of this north-facing photograph. (George Hamlin photo)

Central Station, situated near the present-day intersection of Michigan Ave. and Roosevelt Rd., is widely regarded as one of Chicago's "lost treasures." This stately Victorian landmark, designed by architect Bradford L. Gilbert, boasted a 13-story clock tower and a spacious train shed. Passengers arriving on one of this station's 12 tracks also could easily make connections to suburban trains departing from the adjacent Roosevelt Road Station.

Opened in time for the start of the World's Columbian Exposition in 1893, this station could be reached only via the tracks of the Illinois Central Railroad (IC), the station's owner. These tracks, however, were in high demand. For most of its history, three carriers used the station: the IC and a pair of New York Central subsidiaries (Michigan Central and the Cleveland, Cincinnati, Chicago & St. Louis Railway, which was also known as the "Big Four"). The Chesapeake & Ohio of Indiana also used the station until 1933.

With 58 long-distance trains (excluding the Chicago South Shore & South Bend—"The South Shore Line"—an electric interurban railway operating from Roosevelt Road Station) in 1942, Central Station had an identical number of intercity arrivals and departures as LaSalle St. Station but slightly fewer than Dearborn (62) and North Western stations (64). IC accounted for slightly more than half of these trains (30), with the remainder nearly evenly split between the Big Four and Michigan Central.[12] However, several hundred more Illinois

Central suburban and South Shore trains arrived and departed daily from Roosevelt Road Station. The two depots were linked by a covered (if only lightly used) walkway, and together saw approximately 480 daily trains.[13]

The range of destinations reachable from Central Station, however, was somewhat less than that of Dearborn, LaSalle, and Union stations. Central's trains directly served 23 of the country's 100 largest cities in 1942, substantially fewer than those served from the other stations.[14] Even this number is deceptively high, considering that six of these cities were in the Northeast and could be reached only via Detroit and southern Ontario by

CENTRAL STATION
Michigan Ave. and 11th Pl.

Central Station's head house and clock tower stood virtually unchanged from the time of the station's completion in 1893 until their demise more than 80 years later. This photo, taken on February 24, 1973, shows the station about a year before it was torn down. It had served its last intercity train on March 5, 1972. (Verne Brummel photo)

(Opposite) This map, showing Central Station's passenger train routes on January 1, 1953, illustrates the heavily used routes of New York Central-affiliated lines to Cincinnati, Detroit, Indianapolis and other cities, and routes of the Illinois Central Railroad. Particularly large numbers of passengers traveled on IC's "Main Line of Mid-America" between Chicago and New Orleans and trains to prominent Florida resorts served by IC's *City of Miami and Seminole* in partnership with other carriers. As the map shows, however, Central Station lacked service to the north and had only a single route to the west; thus, it was not a well-balanced hub for transfers between trains.

way of the Michigan Central and New York Central railroads. This route is considerably longer and more time consuming than the more southerly routes serving the same eastern endpoints from LaSalle and Union stations.

Central Station was a powerhouse, however, when it came to serving regional markets. Its trains reached four of the five largest cities within 300 miles of Chicago—Cincinnati, Detroit, Indianapolis, and St. Louis—as well as Grand Rapids, MI, which ranked 10th. Only Union Station served more cities (its trains reached eight of the 10 largest cities) and that rival depot lost its Detroit service in the 1950s. In addition, Central Station had a large share of the rail market to all of these places well into the 1960s, with the exception of Grand Rapids, to which direct service ended the previous decade. Travelers could find no faster service to Cincinnati than the New York Central's *James Whitcomb Riley*, or a faster train to Detroit than Michigan Central's *Twilight Limited*, both of which departed from Central Station.

For longer-distance travelers, Central Station had the dominant route to Memphis and New Orleans—both of which were on IC's well-known "Main Line of Mid-America" used by the *City of New Orleans, Panama Limited*, and other express trains. Warmer climes could also be reached by the *City of Miami* (jointly operated by the IC and several connecting railroads; and departing every other day)—a popular way to travel to the Sunshine State.

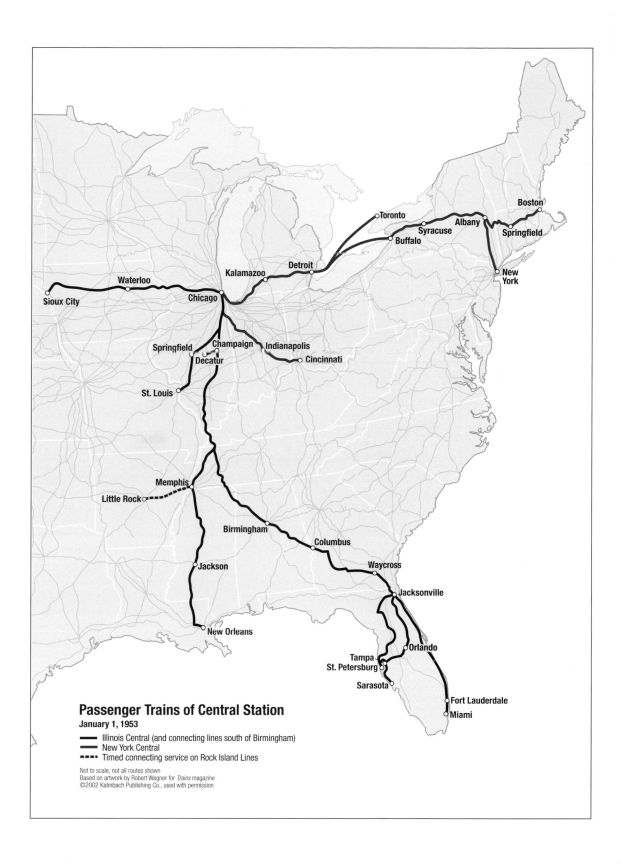

Passenger Trains of Central Station

January 1, 1953

— Illinois Central (and connecting lines south of Birmingham)
— New York Central
- - - - Timed connecting service on Rock Island Lines

Not to scale, not all routes shown
Based on artwork by Robert Wegner for *Trains* magazine
©2002 Kalmbach Publishing Co., used with permission

CENTRAL STATION
Michigan Ave. and 11th Pl.

Viewed through the window of Central Station's control tower on August 15, 1963, New York Central's *James Whitcomb Riley* (with Illinois Central locomotives), and IC's *Land O'Corn* and *Panama Limited* can be seen side by side, awaiting their respective departures. The angular train shed canopy was added in 1945 to replace the taller original structure. Tracks leading up to freight facilities along the Illinois Central's St. Charles Air Line—a connecting route used by Iowa-bound trains—are at the far right. (Richard Gill photo)

Central Station's role as a self-contained connecting hub was nonetheless limited by the fact that almost all trains departed toward the east, southeast, and south, making most connections rather circuitous.[15] The Illinois Central did have a route to Sioux City, IA—known for such trains as the *Hawkeye* and *Land O'Corn*—but this corridor was of relatively minor significance and not used for commuter service after 1931. Most passengers traveling between places on opposite sides of Chicago, consequently, had to change terminals after arriving at Central.

For several years, starting in 1963, the Soo Line operated into the terminal, having stopped using the deteriorating Grand Central Station and giving Central Station direct service to Wisconsin and Michigan's Upper Peninsula.[16] Like so many others, these trains were gone by 1968, and by the middle of the following year, the number of daily long-distance arrivals and departures at Central had dwindled to 18.

When Amtrak assumed responsibility for the remaining intercity service in May 1971, service was pared to three daily pairs of trains. These trains began operating out of Union Station in early 1972, leaving Central's massive waiting room and platforms completely vacant.

The grandeur of the station was gradually compromised, particularly after a false ceiling was put over the vaulted waiting room in 1962, in part to lower the cost of heating the building. Illinois Central moved its headquarters to the Illinois Center complex farther north along its tracks and demolished the station in 1974. Roosevelt Road was extended across the tracks in the late 1990s and much of the old station's grounds were transformed into a grassy lawn that today constitutes the southwest corner of Grant Park.

The electrified former Illinois Central suburban corridor, however, remains heavily used by Metra and the South Shore Line. One hundred and forty four weekday trains arrive and depart from platforms near the historic terminal site. After deteriorating to the point that they were considered an eyesore, the Roosevelt Road Station platforms were demolished in 2009 to make way for Metra's new Museum Campus Station, at which passengers reach the platforms from the opposite end as in the previous station.

Virtually nothing remains of Central Station today except several massive concrete blocks near the corner of Michigan Ave. and Roosevelt Rd., which are said to be the cornerstones of the old terminal. The generous width of the railroad corridor nearby and the imposing retaining walls alongside it are nonetheless vivid reminders of the importance of this lost historical landmark. A new residential neighborhood nearby on the site once partially covered by the station's yards is named "Central Station" in honor of the bygone terminal.■

The canopies, platforms, and tracks below Central Station's main waiting room await the next train, which was destined to never arrive, on July 13, 1974. Aging signs above the platforms inform nonexistent travelers of the station's name and owner. Demolition began soon after this photo was taken. (Verne Brummel photo)

DEARBORN STATION
47 W. Polk St.

Santa Fe's *Grand Canyon* appears ready for a departure from Dearborn Station on June 6, 1962. The locomotives' colorful "war bonnet" paint contrasts sharply with the aging station. This photo was taken from the Roosevelt Rd. overpass, one of the city's most popular train-watching locales. (Bob Krone photo)

Dearborn Station was the elder statesman of Chicago's downtown terminals, being both the oldest and the one serving the greatest number of passenger railroads. A rambling facility located at Dearborn & Polk, this station is known to many contemporary Chicagoans for its distinctive masonry head house and clock tower, which are still standing today. Among transportation enthusiasts, however, it is fondly remembered for hosting some of Chicago's most famous streamliners. Santa Fe's famed *El Capitan*, *Super Chief*, and *San Francisco Chief's* afternoon departures kept Dearborn a relatively busy place until almost the end.

This three-story structure exudes the strong Romanesque Revival tastes of station architects at the time of its construction. Opened in 1885 and built out of pink granite and red brick, it was owned by the Chicago & Western Indiana Railroad (C&WI). With tracks extending south from the station to Hammond, IN, C&WI provided access to downtown Chicago for its own trains and six other carriers: Chicago & Eastern Illinois, Erie, Grand Trunk Western, Monon, Santa Fe, and Wabash. At the turn of the twentieth century, as many as 122 trains (both commuter and intercity runs) used the station daily, offering direct service to hundreds of cities across the country and into eastern Canada.

A massive train shed, measuring 165 feet wide and 700 feet long, was one of the largest in the Midwest, despite its archaic design. The appearance of the head house and tower, designed by architect Cyrus L.W. Eidlitz, were somewhat diminished when their pitched roofs were eliminated (Dearborn became a flat-topped structure after a fire in 1922), but the station remained one of downtown's most prominent landmarks.

In 1942, Dearborn Station hosted 64 long-distance trains, putting it behind only Union Station (124 trains) with respect to intercity activity. But it had six long-distance carriers, the most of any Chicago station. Moreover, its mix of services differed greatly from those of Central, LaSalle St., North Western, and Union stations. Among them, these stations had just one carrier—the Nickel Plate, which used LaSalle Street Station—with eight or fewer daily intercity trains. At Dearborn, all carriers except the Santa Fe had this distinction. Most of Dearborn's tenants, consequently, were smaller players in local rail-travel markets. Another difference: Dearborn's commuter business during World War II never amounted to more than two pairs of weekday trains—roundtrips on the C&WI to south-suburban Dolton, IL, and a Wabash round trip serving the southwest suburbs.

When viewed in its entirety, however, the scope of Dearborn's service stands out. In 1945, the station's trains directly served 38 of the country's 100 largest cities, putting it behind only Union and LaSalle St. stations.[17] Most impressively, Dearborn's trains reached 13 of the country's 25 largest cities, second only to Union Station. Moreover, Dearborn's trains offered direct service to many points that were not accessible from Union Station, including Arizona, Louisiana, Texas, southern California, and eastern Canada.

The Chicago & Eastern Illinois Railroad's *Danville-Chicago Flyer*, led by a locomotive of Louisville & Nashville, a partial owner of the company, stands at the Track 6 bumper post inside Dearborn Station's massive trainshed on December 26, 1969. Grand Trunk Western's *International*, an overnight train from Toronto, ON, operated in conjunction with GTW's owner, Canadian National, occupies an adjacent track. (Marty Bernard photo)

A shuttered Dearborn Station awaits an uncertain fate in early 1976, five years after it had served its final intercity train on the eve of Amtrak's creation. As evidenced by the crane at the right edge of the photo, demolition of the train shed was already under way. The Romanesque Revival clock tower and head house, dating to 1885, were spared and survive as South Loop landmarks. (Craig Bluschke photo)

DEARBORN STATION
47 W. Polk St.

Dearborn Station offered direct service to 21 states and 2 Canadian provinces at the start of 1950. This map shows the full extent of the station's services at the height of the postwar era. Santa Fe's famed *Chiefs*, operating to the Southwest, as well as the Chicago & Eastern Illinois Railroad's *Dixie* trains, which operated in conjunction with connecting railroads to provide direct service to Florida, were among the station's showpieces. The southerly orientation of Dearborn Station's routes and the Erie Railroad's status as a relatively minor player in the Chicago–East Coast market, however, limited this terminal's role as a self-contained transfer point for long-distance travelers.

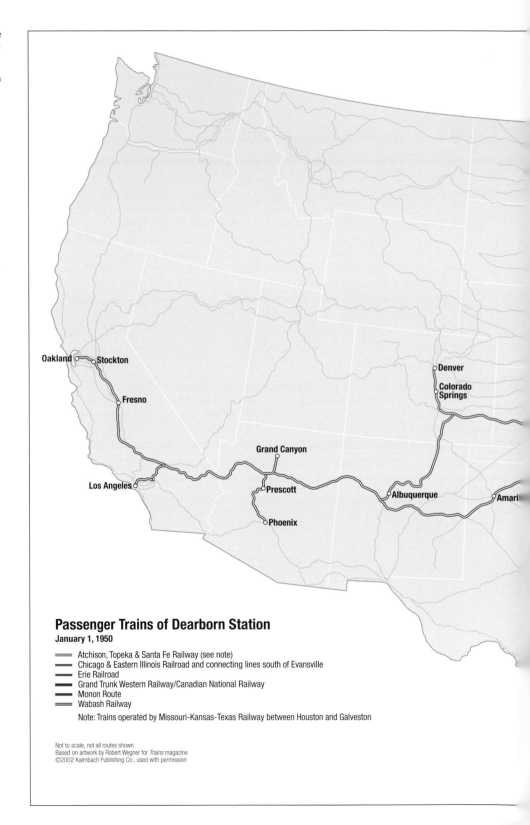

Passenger Trains of Dearborn Station
January 1, 1950

- Atchison, Topeka & Santa Fe Railway (see note)
- Chicago & Eastern Illinois Railroad and connecting lines south of Evansville
- Erie Railroad
- Grand Trunk Western Railway/Canadian National Railway
- Monon Route
- Wabash Railway

Note: Trains operated by Missouri-Kansas-Texas Railway between Houston and Galveston

Not to scale, not all routes shown
Based on artwork by Robert Wegner for *Trains* magazine
©2002 Kalmbach Publishing Co., used with permission

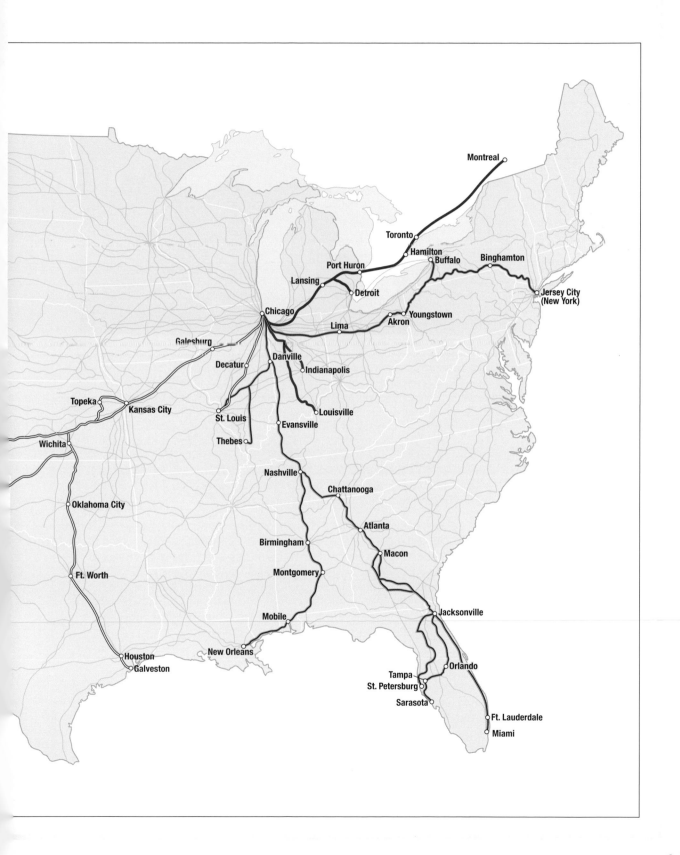

DEARBORN STATION
47 W. Polk St.

Because of its diverse mix of tenant railroads, Dearborn also was an efficient, self-contained connecting hub, particularly between points northeast and southeast of Chicago and to the American Southwest. Only Union Station excelled to a greater extent as a connecting complex. The Chicago & Eastern Illinois "Dixie" trains, operated with several connecting lines, were a popular way to reach the Deep South and Florida, while the station's powerhouse, the Santa Fe, offered the fastest service available to Houston and Oklahoma City (via the *Texas Chief*), Los Angeles (*Super Chief* and *El Capitan*), and Phoenix. Even so, Dearborn's services to the Eastern Seaboard were limited to a handful of Erie Railroad trains, a minor player in this market.

Travelers took a strong liking to the Santa Fe's *Chiefs*, allowing Dearborn to more slowly surrender its transportation role than most of the other city terminals. The number of long-distance trains dropped 13 percent, to 56 daily, between 1942 and 1956, the lowest proportional drop among the six downtown stations. The waiting room was redesigned and ticketing areas freshened up to improve the station's ambiance. In 1954, one of the last new streamliners to be inaugurated in the country—Santa Fe's *San Francisco Chief*—rolled out of this station for the first time.

The period from 1956 to 1969 was less favorable to Dearborn, spelling doom to almost three-quarters of its long-distance trains and its commuter trains to Dolton. Although this aging station had lost only one carrier, the Monon, and thus still had more *carriers* than any Chicago station, its long-distance service at the end of this period totaled just 16 daily trains. All intercity service from Dearborn ended in 1971, with the startup of Amtrak, when Santa Fe trains were rerouted to Union Station by the new national carrier. Norfolk & Western still operated its former Wabash Railway commuter service to Orland Park, IL, from a platform adjacent to Dearborn's main terminal building, but the station itself was closed. In 1976, this N&W operation was relocated to Union Station, clearing the way for property redevelopment on the Dearborn site.

A modernization project in the late 1940s gave Dearborn Station improved ground-floor ticketing and baggage-handling facilities as well as a second-floor waiting room that overlooked arriving and departing trains, all of which are visible in this 1950s-era photo. A large crowd can be seen congregating in the gate area, likely for the impending departure of one of Santa Fe's famed *Chiefs*. (Santa Fe collection, Kansas State Historical Society)

New signs and lockers (at right), restaurant facilities (left), and a glass-protected arrivals and departures board (top) installed in the late 1940s improved Dearborn Station's image among travelers. Almost all evidence of the terminal's late-nineteenth century interior was eliminated. The clock shows 1:30—the rush of travelers departing aboard the station's popular overnight trains to the southern and western United States tended to reach its peak later in the afternoon. (Santa Fe collection, Kansas State Historical Society)

As the threat of demolition loomed, city officials explored possible reuses of the disused station's train shed. Some considered the historical preservation of the deteriorating shed as more important than the station's head house. Nevertheless, the train shed was demolished in 1976, and the head house was spared. Much of the area once covered with the station's platforms was transformed into the Dearborn Park residential complex.

DEARBORN STATION
47 W. Polk St.

A day after Christmas, on December 26, 1967, the west side of Dearborn Station's ambling train shed is a busy place. A Grand Trunk Western passenger train sits behind the lengthy Santa Fe streamliner at center, which is likely the *Super Chief/El Capitan* and has a full length dome "deadheading" in front of its baggage cars. The trusses and columns looming overhead are defining features of the station's cavernous interior. (Marty Bernard photo)

Today, Dearborn Park is credited with helping the revival of the South Loop. The head house is a cherished landmark, standing watch over Dearborn Park and the Printers Row historic district. Visitors walking south down Dearborn St. can see the clock tower looming in the distance—a testament to the depot's role in the city's development. Those using the Polk St. exit from the CTA's Harrison Red Line stop can still find a mosaic wayfinding sign, intended to facilitate passenger transfers, pointing the way to a station that has not seen a train in more than 40 years. ∎

The gabled mail and express building on the west side of Dearborn Station glows in the sun on July 13, 1974. A portion of the station's train shed can be seen at the upper right. The differing times on the clock faces attest to the fact the station is closed to the public. (Verne Brummel photo)

(Opposite) Dearborn Station's 1888 clock tower is partially illuminated by the evening sun in 2012. Because traffic on Dearborn moves toward the north, many motorists see the terminal only through their rearview mirror, so this landmark remains somewhat less appreciated by the public than other historical downtown buildings. (Eric Allix Rogers)

GRAND CENTRAL STATION
201 W. Harrison St.

Members of the Chicago Art Club paint scenes of Grand Central Station from the grassy median of the new Congress Expressway on August 11, 1960. The station's clock tower and dramatic billboard promoting trains of owner Baltimore & Ohio made this an enticing vantage point. Impeccable service aboard the three advertised trains ensured that Grand Central remained dominant in the market from Chicago to Baltimore and Washington well into the 1960s. (Author's collection)

Grand Central Station, located at Harrison and Wells streets, was a towering legacy of the ambitions and aspirations of railroad leaders at the height of Chicago's Golden Age. This station's arched train shed and head house in the Norman castellated style, with its 16-story clock tower, gave it a majestic presence. The train shed, an ornate iron structure built for a subsidiary of the Northern Pacific Railway in 1891, was the second largest in the world at the time of its construction, exceeded in width only by St. Louis Union Station.

This station, sold in 1910 to Baltimore & Ohio, was used by four railroads from 1940 through the 1960s: B&O, Chicago Great Western, Pere Marquette (after 1947, part of Chesapeake & Ohio), and Soo Line (during and after World War II). Grand Central's role in Chicago-area transportation nonetheless pales in comparison to that of its illustrious neighbors—LaSalle Street Station to the east and Union Station to the west. Grand Central served fewer passengers than any of the other five downtown railroad stations throughout the early twentieth century, but the B&O's long-distance trains made it a dynamic part of the downtown travel scene.

Trains leaving Grand Central diverged from the metropolitan region on four different routes in 1942, but the level of service to the east was much stronger than to the north, south, or west. Passengers could travel directly to 20 of America's 100 largest cities, but several (including Philadelphia, New York, and Wilmington, DE) were accessible only via B&O trains on a circuitous route north from Washington, D.C. Travelers had a choice of carriers to only one metropolitan area—the Twin Cities, served by Chicago Great Western (via Oelwein, IA) and Soo Line. Both of these carriers, however, were minor players on this highly competitive route.

B&O's trains from Chicago to Washington, D.C., most notably the *Capitol Limited* and *Columbian*, were the station's most famous, with the latter providing the fastest service available to both Akron and Youngstown, OH, and—more significantly—the nation's capital. Both also offered highly competitive service to Pittsburgh and Baltimore, their final destination. Additionally, Soo Line was one of just two carriers operating direct service to Duluth, MN. On shorter routes, though, Grand Central was less

Carts stacked with express shipments next to a pair of baggage cars with wide-open doors illustrate the importance of such "head-end" (front of train) business to the railroads as passenger levels declined after World War II. The cars seem small inside Grand Central's soaring train shed, which measured 156 feet wide and 555 feet long, making it one of largest in the world at the time of its construction in 1890. This view faces north towards the glass-paneled wall separating the trainshed from the head house. (Historic American Building Survey, Library of Congress)

consequential, with its trains directly serving just one of the 10 largest cities (Grand Rapids, MI.) within 300 miles of Chicago. The *Pere Marquette* streamliner had the lion's share of the Grand Rapids business, but service between Grand Central and most other regional markets was generally weak.

Many remember Grand Central as a relatively uncongested place. It had 28 long-distance trains in 1942—less than half the number of any of the other "Big Six." The number of trains dropped 29 percent, to 20 daily, between 1942 and 1956—the greatest share of any of the city's downtown stations. Chicago Great Western ended passenger service in 1956, and Soo Line moved its remaining trains to Central Station in 1965. All service at Grand Central ended on November 8, 1969, when B&O and C&O (by that time corporate affiliates) relocated their remaining trains to North Western Terminal.

Civic leaders expressed hope that Grand Central would be declared a national historic landmark. Concerned that this status would impede its ability to sell the property, B&O had it demolished in 1971, just as Chicago's historic-preservation movement was beginning to gain traction. Almost immediately, the civic community bemoaned the loss of this architectural treasure. Some have speculated that, had it stood for a few more years, the train shed might have been targeted for reuse in a way similar to the old Reading Terminal in Philadelphia, in which the train shed was transformed into a convention center ballroom.

Much of the land south of the station was transformed into the River City apartment and condominium complex, which opened in 1986. Most of the former station grounds, however, still await redevelopment. Numerous proposals to reuse the property, which remains in the hands of a private developer, have foundered.

GRAND CENTRAL STATION
201 W. Harrison St.

One of Grand Central Station's most distinguished and longest-serving trains, B&O's *Capitol Limited* awaits its afternoon departure in July 1966. A train of B&O affiliate Chesapeake & Ohio, mostly likely the daily *Pere Marquette* to Grand Rapids, MI, is visible in the distance, to the left of the Washington-bound streamliner. (George H. Drury photo)

(Opposite top) Grand Central Station's passenger-train routes, shown here on January 1, 1950, radiated in four directions, with Baltimore & Ohio's Chicago–Washington, D.C. main line being its most prominent and heavily used travel artery, offering premier service to the nation's capital. B&O trains offered convenient through-train service to Baltimore, Philadelphia, and metropolitan New York. The Pere Marquette Railway (owned since 1947 by Chesapeake & Ohio) dominated the rail-passenger business to Grand Rapids, MI, and offered well-timed connections to Detroit, albeit over a relatively circuitous route

(Opposite bottom) Little remains of Grand Central Station beyond these remnants of its concrete platforms. Many proposals to redevelop this parcel of land have come and gone over the years. This photo was taken from a vantage point similar to the 1966 view of B&O's *Capitol Limited*. (Author's photo)

Several of Grand Central's platforms survive, but are crumbling in a grassy fenced-in area often used by South Loop residents as a dog park. Those visiting the site are often struck by the presence of this vacant land in close proximity to the Chicago Board of Trade and Willis Tower, commercial icons only a few blocks away. ■

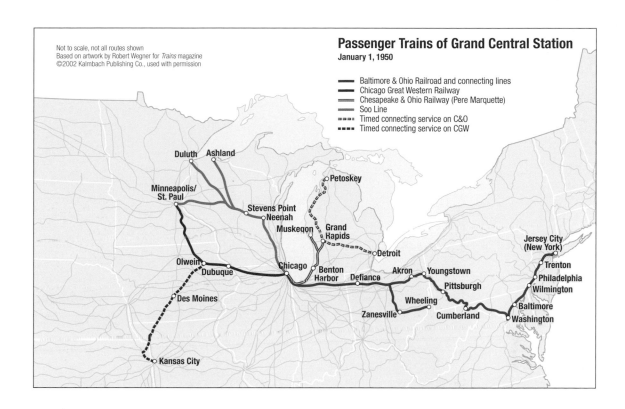

Passenger Trains of Grand Central Station
January 1, 1950

—— Baltimore & Ohio Railroad and connecting lines
—— Chicago Great Western Railway
—— Chesapeake & Ohio Railway (Pere Marquette)
—— Soo Line
==== Timed connecting service on C&O
■■■■ Timed connecting service on CGW

Duluth Ashland
Minneapolis/
St. Paul
Stevens Point
Neenah
Muskegon
Grand
Rapids
Petoskey
Detroit
Olwein
Dubuque
Chicago
Benton
Harbor
Defiance
Akron
Youngstown
Pittsburgh
Des Moines
Wheeling
Zanesville
Cumberland
Jersey City
(New York)
Trenton
Philadelphia
Wilmington
Baltimore
Washington
Kansas City

LaSALLE STREET STATION
414 S. LaSalle St.

LaSalle Street Station's main entrance was largely concealed by the Van Buren St. "L" structure. This photo, taken from the LaSalle "L" station on October 26, 1974, shows LaSalle Street Station's graceful arched entrance from one of few vantage points offering an unobstructed view. The station, with its "head house" functions occupying the lower floors of a 13-story office building, opened in July 1903. (Craig Bluschke)

LaSalle Street Station is deeply rooted in modern passenger-train history as the primary downtown terminal for the New York Central (NYC) and Chicago, Rock Island & Pacific (Rock Island) railroads—two of the city's most important long-distance carriers—as well as the New York, Chicago and St. Louis Railroad (Nickel Plate Road). Located at the intersection of LaSalle and Van Buren streets, it is remembered for the *20th Century Limited*, an all-Pullman train to New York; the *Golden State*, a popular run to California; and other express trains. The station's head house, designed by Frost & Granger, has been demolished, but the modernized facility remains a heavily used commuter-rail terminal.

LaSalle Street Station's vaulted main waiting room is decorated for the Christmas season in this November 1981 photo. Passengers could access the platforms through the doors at center left. Within a few months, the station building had been unceremoniously demolished to make way for redevelopment. (Craig Bluschke photo)

Opened in 1903 to replace an older station with the same name, LaSalle Street Station had tracks elevated above the street to eliminate grade crossings at the terminal's throat. With trains operating as far east as Boston and as far west as Los Angeles, it was a major player in intercity travel, particularly on east-west routes. The station had 58 long-distance trains in 1942, tying it with Central Station as the city's fourth busiest. LaSalle Street's trains directly served 32 of the country's 100 largest cities in 1942, putting it behind only Dearborn and Union stations on this measure. Approximately 100 commuter trains also used the stations at the time.

Passengers making connections in Chicago benefited from LaSalle's central location. The station had a direct connection to the LaSalle "L" stop and was only a block from the city's bustling financial district. Such convenience contributed to the enormous popularity of New York Central's *20th Century Limited*, once heralded by the carrier as "the Most Famous Train in the World." The number of passengers making transfers to other trains from LaSalle, however, was limited by the convenience of Englewood Union Station, eight miles south of LaSalle. Passengers of all three carriers serving LaSalle (as well as those of the Pennsylvania Railroad, which operated from

These staircases linked LaSalle Street Station's lower-level entrance with the upper-level waiting room, ticket windows, and baggage-check area. The upper floor also offered direct access to the outer platform of the Loop "L"—a popular link for commuters arriving on Rock Island Railroad trains and needing to reach jobs more than a few blocks away. Note the streamlined appearance of some of the station's modernized interior surfaces. (Craig Bluschke photo)

LaSALLE STREET STATION
414 S. LaSalle St.

As is evident on this map of LaSalle Street Station's trains as of January 1, 1950, the centrally located terminal was an efficient transfer point for passengers on long-distance trips, particularly those having an east–west orientation. The station benefited from the complementary nature of the New York Central and Rock Island Lines systems. The Nickel Plate Road provided competition for NYC to Cleveland and Buffalo and offered through-car service to Hoboken, NJ, jointly with the Lackawanna Railroad. NYC was a dominant player in the Boston, Buffalo, Cleveland, and New York markets and served Pittsburgh in partnership with the Pittsburgh & Lake Erie Railroad.

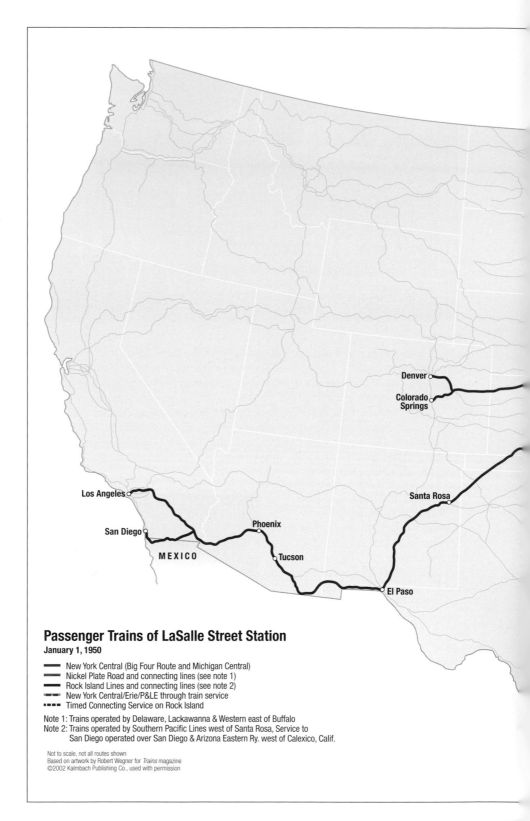

Passenger Trains of LaSalle Street Station
January 1, 1950

━━━ New York Central (Big Four Route and Michigan Central)
━━━ Nickel Plate Road and connecting lines (see note 1)
━━━ Rock Island Lines and connecting lines (see note 2)
━━━ New York Central/Erie/P&LE through train service
▪▪▪▪ Timed Connecting Service on Rock Island

Note 1: Trains operated by Delaware, Lackawanna & Western east of Buffalo
Note 2: Trains operated by Southern Pacific Lines west of Santa Rosa, Service to
      San Diego operated over San Diego & Arizona Eastern Ry. west of Calexico, Calif.

Not to scale, not all routes shown
Based on artwork by Robert Wegner for *Trains* magazine
©2002 Kalmbach Publishing Co., used with permission

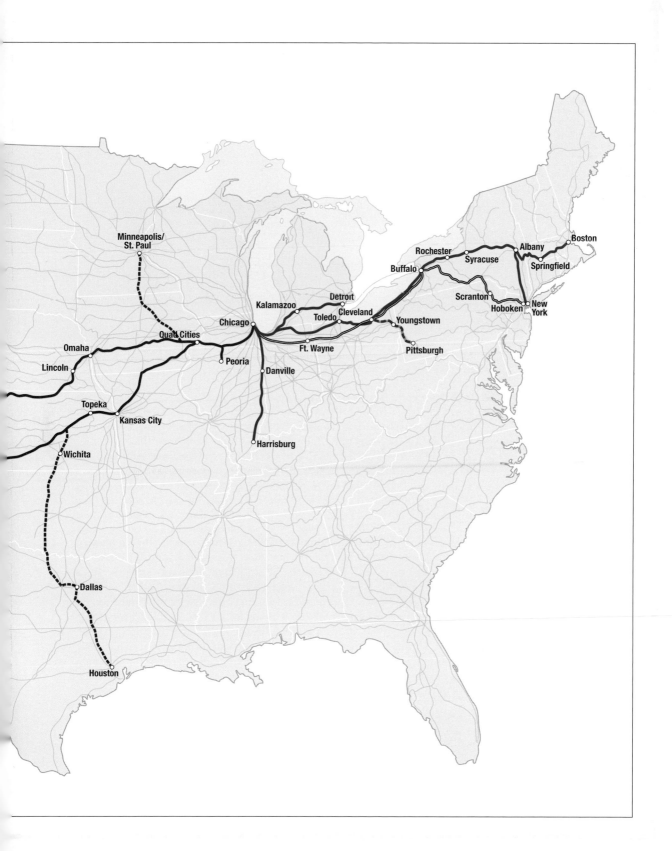

LaSalle Street Station
414 S. LaSalle St.

LaSalle Street Station's head house and train shed are somewhat idealized in this 1930s-era postcard. The artist embellished some of the station's features, including the amount of space between the terminal's main entrance and the "L," and also enhanced the grandeur of the train shed. Nevertheless, the image is a reasonable depiction of the station's architecture and layout.
(Author's collection)

110   LA SALLE STREET STATION, CHICAGO

Union Station) could make connections at Englewood, thereby eliminating the need to travel to downtown Chicago to change trains.

The Rock Island and New York Central (NYC) complemented each other well, operating trains in the early postwar years linking 13 and 16 of the top 100 cities, respectively, with no competitive overlap. New York Central was a preferred choice for travel to Buffalo, Cleveland, and New York City, and offered the only direct service available to Rochester, Syracuse, and Albany, NY, as well as Boston. Rock Island (which maintained its company headquarters in the terminal building) boasted the fastest service available to Des Moines and the Quad Cities, with highly competitive service to Kansas City and (in conjunction with Southern Pacific) to El Paso, Phoenix, Los Angeles, and San Diego.

LaSalle Street Station, nonetheless, had notable weaknesses as a self-contained transportation hub. The Nickel Plate's trains directly served just eight of the country's top 100 cities, and all except two (Fort Wayne, IN, and Scranton, PA) could be reached on the New York Central in less time, limiting Nickel Plate's importance to the Chicago travel market. Moreover, LaSalle Street tended to be weak precisely where Central Station was strong, as it lacked direct service to most major cities in the Midwest and Southeast, including Cincinnati, Indianapolis, Florida, and the Deep South. LaSalle had direct service to just three of the 10 largest population centers (Detroit, Toledo, and Quad Cities) within 300 miles of Chicago, whereas Central and Union stations had service to five and eight of these cities, respectively.

Despite this, LaSalle lost a smaller share of its trains between 1942 and 1956—just 17 percent—than any of the "Big Six" stations, with the exception of Dearborn. Both New York Central and Rock Island were deeply committed to the passenger business after the war and invested heavily in new streamlined cars and station improvements. NYC also began operating all of its Detroit trains, most of which had previously used Central Station, to and from LaSalle Street.

Starting in the mid-1960s, however, cutbacks had devastating effects. Passenger service on the former Nickel Plate ended in 1965 (a year after the merger with Norfolk & Western). Rock Island's commuter business suffered as droves of downtown workers abandoned its trains in favor of driving on the newly constructed Dan Ryan Expressway, or taking the rapid-transit line in its median. New York Central's *20th Century Limited*, was downgraded and stripped of its name in 1967. Service to California ended when the *Golden State*, a Rock Island train jointly operated with Southern Pacific, was dropped in 1968.

The bottom fell out later that year when New York Central's successor, Penn Central (the result of NYC's merger with the Pennsylvania Railroad), transferred its entire LaSalle operation to Union Station, reducing LaSalle's long-distance traffic to a mere six trains by 1969. Its remaining intercity service could have ended upon creation of Amtrak, as at Dearborn Station and North Western Terminal, had the Rock Island not remained independent of this newly formed, quasi-governmental entity. Instead, Rock Island was compelled to continue operating trains to Peoria and the Quad Cities for several more years before mercifully being permitted to discontinue them in 1978. When these trains were annulled, LaSalle Street Station became a commuter-only facility.

The station's head house was demolished in 1981 to permit construction of the Midwest Stock Exchange and One Financial Place office tower. In the process, the rail terminus was moved a block farther south and the number of tracks reduced from 12 to eight. The redesigned facility was criticized for lacking a traditional front door and convenient access to rapid-transit lines. Although the Blue Line passes directly underneath the station, there is no direct entrance. Metra, the owner of the new LaSalle Street Station, made gradual improvements—most notably, a massive canopy over the passenger concourse.

A neon sign near LaSalle Street Station's main entrance points travelers to the ticket window, circa 1972. The words "New York Central" on the right ride of the sign are no longer lit, since Penn Central had shifted its former NYC schedules to Union Station in 1968. (Mark Llanuza photo)

The ironwork above the largely empty platforms at LaSalle Street Station on April 9, 1978 evokes memories of busier times, when trains of three railroads arrived and departed here. A bi-level Rock Island commuter train is closest to the camera, while the adjacent train is a special run led by the Rock Island locomotive painted to commemorate the U.S. bicentennial in 1976. (Craig Bluschke photo)

LaSALLE STREET STATION
414 S. LaSalle St.

A brightly painted Rock Island passenger diesel peers out of LaSalle Street Station's train shed in April 1971. Because Rock Island opted not to join Amtrak for financial reasons, its limited schedule of intercity *Rocket* trains serving the Quad Cities and Peoria continued to operate out of LaSalle Street Station until December 31, 1978. The station's head house is visible through the metal trusses supporting the train shed's roof. Although Metra commuter trains use a modernized station on this site, the head house and train shed have been demolished. (George Hamlin photo)

LaSalle Street Station today sees 35,000 weekday travelers boarding and disembarking 68 trains on Metra's Rock Island District. The station appears to have a bright future. In 2011 the city opened the LaSalle–Congress Intermodal Transfer Center, a CTA bus terminal, just west of LaSalle's platforms and south of Congress Parkway. This improvement greatly improved passenger flow and bus-rail connections. To address growing congestion at Union Station, future plans call for relocating some of that station's commuter trains—Metra's Southwest Service route—to LaSalle Street Station. ■

On a chilly and rainy evening in 1975, the *Quad Cities Rocket*—at the time, one of LaSalle Street Station's two remaining long-distance trains—waits at the platform, with distinctive dome-observation car *Big Ben* at the rear. Steam escaping from the cars' heating system contributes to a gloomy ambiance. A commuter train stands two tracks over. (Mark Llanuza photo)

New York Central's *20th Century Limited*, having arrived from New York City, is about to be pulled to the yard by a switch engine in September 1959. The external appearance of this famous train suggests that it is still in fine condition, more than a decade after it was reequipped with streamlined cars. LaSalle St. Station's red-brick head house and the Chicago Board of Trade Building are visible in the distance. (John Dziobko photo)

NORTH WESTERN TERMINAL
500 W. Madison St.

The evening sun illuminates North Western Terminal's Italianate facade in December 1976. The station's majestic appearance belied its diminishing status, relegated to commuter service only. The main entrance on Madison St. (left center) had already been blocked in favor of a more modest access point below the overhead footbridge crossing Canal St. (right)—the most popular way to reach the station. (Craig Bluschke photo)

North Western Terminal's enormous dimensions and ornate Renaissance Revival colonnade were unmistakable reminders of the importance of the Chicago & North Western Railway (C&NW) to transportation in its home city. Through World War II and into the early postwar era, C&NW operated more intercity trains from this station than any other carrier did from a downtown station in Chicago.

Opened in 1912 to replace C&NW's Water Street Station (a facility on the site of today's Merchandise Mart), the magnificent North Western Terminal took up an enormous swath of land between Canal and Clinton streets north of Madison. With 16 tracks elevated above street level, two passenger concourses, and a massive but elegant waiting room (known for its barrel-vaulted ceiling), the station was built to serve great passenger volumes in style and comfort. This terminal, designed by the noted architectural firm Frost and Granger, was second only to Chicago Union Station with respect to its train-handling capability and retail businesses.

The station experienced a surge in passenger traffic during World War II, with 182 weekday commuter trains and 62 intercity departures in 1942 pushing capacity to the limit. As the station's name implies, these trains principally served points north and west of the city. Because of the relatively sparse populations of the regions served by C&NW, the terminal's trains reached just 15 of the country's 100 largest cities in 1942, by far the smallest number served by any of the six downtown terminals. C&NW directly served five of these cities, with the remainder accessible on the *Challenger, City of Portland*, and other trains operated jointly with its Union Pacific interchange at Omaha, NE.

Still, North Western Terminal served an enormously diverse spectrum of small and medium-sized towns. Moreover, this was the only Chicago station that boasted direct service to the three largest West Coast markets—Los Angeles, the San Francisco Bay Area, and Seattle–Tacoma—throughout World War II and the early postwar years. It also offered the only direct service to Boise, ID, Cheyenne, WY, and Las Vegas, NV, as well as the fastest service available to Salt Lake City.

Another notable aspect of this terminal's history is the sheer number of intercity trains that were operated by C&NW in 1942. With 62 trains, C&NW had the largest intercity presence of any carrier at a downtown Chicago terminal. Passengers flocked to the carrier's fast "400 series" streamliners operating on express schedules to some of its most prominent destinations. The *Twin Cities 400*, which gave birth to the "400" brand, covered the 400 miles to St. Paul, MN in an equal number of minutes. The *Dakota 400* ran to Rapid City, SD, the *Kate Shelley 400* to Clinton, IA, and the *Shoreline 400* to Green Bay, WI. Summer vacationers took a particular

A multiplicity of routes converging on North Western Terminal, depicted here as of January 1, 1957, made this downtown station a major player for passenger travel to Minnesota, Wisconsin, and Michigan's Upper Peninsula. The station's trains by this point were confined to the Upper Midwest and Great Plains states as the station's owner, Chicago & North Western Railway, no longer handled Union Pacific trains east of Omaha (these trains operated over the rival Milwaukee Road into Chicago Union Station starting in 1955). C&NW's famed "400" stable of trains provided formidable competition to the Milwaukee Road *Hiawathas* and Burlington Route *Zephyrs*.

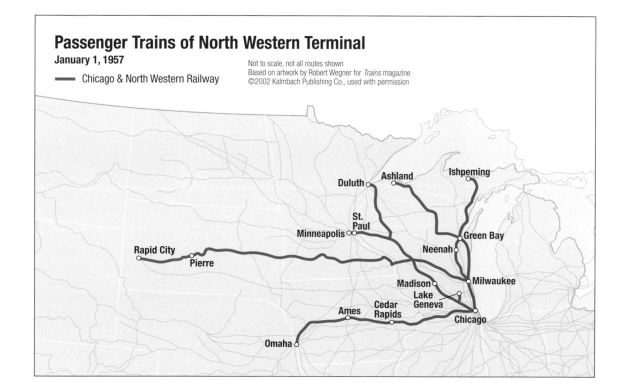

Passenger Trains of North Western Terminal
January 1, 1957

—— Chicago & North Western Railway

Not to scale, not all routes shown
Based on artwork by Robert Wegner for *Trains* magazine
©2002 Kalmbach Publishing Co., used with permission

NORTH WESTERN TERMINAL
500 W. Madison St.

Chicago & North Western timetable from September 1939 promotes service on Union Pacific to the Pacific Coast.

C&NW commuter trains, with their familiar yellow-and-green paint scheme, cluster at the north end of North Western Terminal, with the CTA Lake St. "L" passing overhead, in 1975. Passengers could transfer between the two different systems using the Northwest Passage, a covered walkway extending from the station's westernmost platform.
(Mark Llanuza photo)

NORTH WESTERN TERMINAL
500 W. Madison St.

Five trains and North Western Terminal's aging but still graceful train shed are visible in this photo from February 13, 1976. Within a few years, most of the trains were repainted in the Regional Transportation Authority's colors and the historic head house, visible in the distance, was no more. (Verne Brummel photo)

liking to the *Ashland 400* to the Lake Superior port near the northern tip of Wisconsin after which it was named. These trains and others also gave North Western Terminal the most frequent service to Milwaukee.

When the carrier terminated its agreement to handle Union Pacific's Chicago-bound trains east of Omaha in 1955, the terminal became the first of the city's six downtown stations to lose all service to points outside the Midwest and Upper Great Plains region. Yet C&NW remained committed to the passenger business—both intercity and commuter—investing in gallery cars with seating on two levels, and improving schedules well into the 1950s. For the following 20 years, its commuter service remained robust and recovered most of its operating costs.

The decline in long-distance service from this terminal accelerated by the early 1960s, a painful process that spelled doom for the *Twin Cities 400* in 1963. The decline was interrupted only by the relocation of B&O and C&O (formerly Pere Marquette) trains from Grand Central Station to this station as part of cost-cutting moves in 1969. This gave North Western Terminal direct service to points east of Chicago for the first time. However, this arrangement proved to be short-lived, and all of the remaining long-distance services, including the once widely used Chicago–Milwaukee trains, stopped when Amtrak began operating on May 1, 1971.

The cutbacks reduced North Western Terminal to a mere commuter station. The two daily round-trips to Lake Geneva, WI, that survived may have had the semblance of a long-distance service, but they were classified as commuter runs, which spared them from elimination upon the creation of Amtrak. These trains lasted only a few more years, making their last runs in 1975.

North Western Terminal's head house was demolished in 1983 and replaced by a modern 32-story office building with an attractive station facility—today's Ogilvie Transportation Center—on its lower levels. The new station features a dramatic glass-enclosed atrium designed by architect Helmut Jahn and a spacious food court popular among travelers and non-travelers alike. In 1991, the train shed and platforms, and its street-level suburban concourse, were totally replaced, but the viaduct remains a vivid testimonial to the glory years of C&NW passenger service. Currently, with 185 daily commuter trains serving 80,000 daily boardings and alightings, Ogilvie is second only to Union Station in daily passenger traffic. ■

North Western Terminal picked up two new tenants—Baltimore & Ohio and Chesapeake & Ohio—shortly before the closure of Grand Central Station on December 31, 1969. Although the B&O and C&O main lines both departed Chicago on an easterly alignment, the move to North Western Terminal meant a circuitous routing that initially headed northwest. This photo shows B&O's recently arrived *Capitol Limited* at the bumper post on December 29, 1969, next to a Chicago & North Western commuter train equipped with bi-level gallery cars. (Marty Bernard photo)

NORTH WESTERN TERMINAL
500 W. Madison St.

The architectural nuance and retail amenities of North Western Terminal stand out in this January 9, 1976, photograph of the main passenger concourse. Escalators in the distance led down to one of the station's Canal St. entrances, while the glass-paneled gates, visible at left, led directly to the platforms. (Verne Brummel photo)

The soaring architecture of Citigroup Center (above), and especially the atrium at Ogilvie Transportation Center, which is located on the second floor, have helped compensate for the sense of loss felt by the demolition of the Chicago & North Western passenger terminal. This view shows the main entrance on Madison St. at an usually quiet time on January 6, 2013—an afternoon in which temperatures fell to -10°F. (Author's photo)

The vaulted ceiling of North Western Terminal's waiting room is aglow from a combination of midday sun and floodlighting, while the floor reflects the fluorescent lights of retail kiosks. The station clock shows 1:30 p.m., well before the evening rush. The terminal's Madison St. colonnade is on the left (south) side on January 9, 1975. (Craig Bluschke photo)

CHICAGO UNION STATION
210 S. Canal St.

In this view from 1970, Chicago Union Station's head house had prominent rooftop signs advertising its four host railroads: co-owners Penn Central and Burlington Northern (Burlington Route prior to March 1970) and tenants Gulf, Mobile & Ohio and Milwaukee Road. Demolition of a smaller concourse building on the east side of Canal St. in 1969 opened up this vista of the station's east façade. (Verne Brummel photo)

Chicago Union Station has for more than a half century offered the greatest range of connecting opportunities for railroad passengers of any station in America. Completed in 1925 to replace an earlier facility with the same name, it features a unique "double stub-end" layout, with tracks leading to the terminal from both the north and south.

Designed by the architectural firm Graham, Anderson, Probst & White, successor to the company created by famed architect and planner Daniel Burnham, this massive 24-track station is unlike any other terminal in the world. Its double stub-end design effectively makes it two stations combined into one. (Only two tracks run through the station, and only one of which can be used by passenger trains). The station's prominent features once included a glass-enclosed concourse (later replaced with a more austere public space below street level) and a large head house with an eight-story office building known for its striking Great Hall.

Union Station served four railroads throughout much of its history: Alton Railroad (later part of the Gulf, Mobile & Ohio); Chicago, Burlington & Quincy (Burlington Route); Chicago, Milwaukee, St. Paul & Pacific (Milwaukee Road); and Pennsylvania Railroad (PRR). In 1942, 124 long-distance trains operated into and out of the terminal, making it the city's busiest intercity station. The massive Beaux-Arts complex had nearly twice as many long-distance trains as Dearborn Station (the city's second busiest station, with 64 trains), and trailed only the great union stations in St. Louis and Kansas City in the number of intercity arrivals and departures among Midwestern terminals. Rush hour was a particularly busy time, with 122 commuter trains also operating to and from Chicago Union Station.

All four railroads serving the facility had a major presence. In 1942, these carriers provided direct service to 43 of the country's 100 largest cities—more than any other U.S. railroad station except St. Louis Union Station and, interestingly enough, the tiny Englewood Union Station on

Rays of light entering Chicago Union Station's main waiting room through its ornate windowpanes create a cathedral-like appearance in March 1943. Two police officers confer in a pool of sunlight near a public-service placard—"Don't Waste Transportation"—related to the war effort. A neon sign above the doorway in the background points the way to limousines of the Parmelee Co., which offered connecting passengers a means of transferring between Chicago's stations—a convenience typically included in the price of their rail tickets. (Jack Delano photo, Office of War Information, Library of Congress)

Flags of the Allied nations hung above passengers using the south end of the concourse at Chicago Union Station in September 1943. The concourse building was connected to the station's main waiting room via a passageway beneath Canal St. Train platforms serving the Burlington Route and Pennsylvania Railroad are at left. A careful study of this photo reveals a sign for the *Zephyr* on the gate at the far left, denoting Burlington's famous fleet of stainless-steel streamlined trains. (Jack Delano photo, Office of War Information, Library of Congress)

Chicago's South Side. Union Station's trains reached 21 of the country's 30 largest cities, a number exceeded only by St. Louis.

In some ways, however, Union Station during this era had more in common with the great terminals of Boston, New York, and Philadelphia than with its Midwestern counterparts. Among the five busiest long-distance railroad terminals in the Midwest (including Cincinnati, Kansas City, St. Paul, MN, and St. Louis), it stood apart for having an extensive commuter-train network in addition to its many long-distance trains. Another notable difference was the limited number of carriers. Union Station had the fewest carriers among the eight stations in the Midwest, with more than 70 daily long-distance trains. The union stations in Kansas City and St. Louis had 16 and 22 carriers, respectively, far more than the four carriers operating out of Chicago Union Station. In this respect, Chicago Union Station had greater similarity to New York's Grand Central and Pennsylvania stations, which had just two and three long-distance carriers, respectively, than it had to its regional peers.

CHICAGO UNION STATION
210 S. Canal St.

The labyrinth of routes from Chicago Union Station, which took passengers directly to more U.S. cities than from any other railroad station in the country, is vividly portrayed on this map, as of January 1, 1946. The serpentine networks of the Burlington Route and its connecting lines, Milwaukee Road, and Pennsylvania Railroad, together with the Alton Route service to St. Louis, gave travelers a dazzling array of choices. Dashed lines depict the western routes added in 1955 when Union Pacific trains began traveling to Chicago over the Milwaukee Road, thereby expanding the reach of Union Station's trains to a remarkable 29 states and the District of Columbia. Several gaps, particularly to Louisiana, New England, and Texas and the Southwest, nevertheless stand out. Several of these gaps were filled after the start-up of Amtrak in 1971, but by then many other routes had been eliminated.

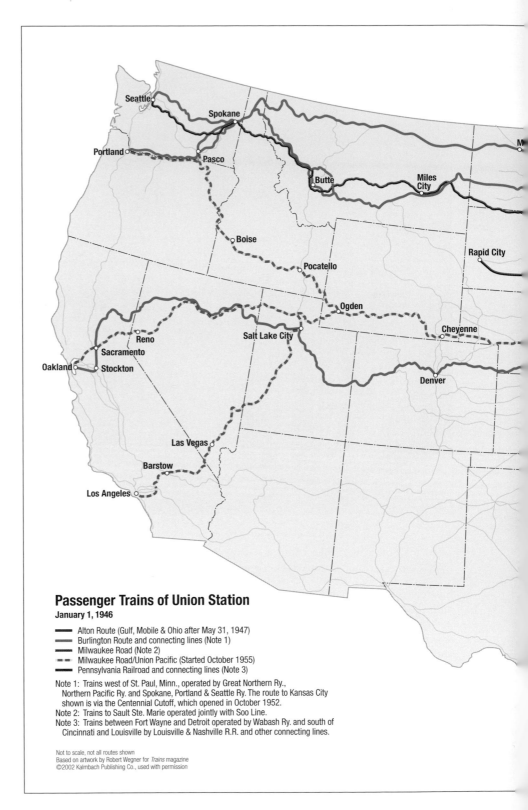

Passenger Trains of Union Station
January 1, 1946

- Alton Route (Gulf, Mobile & Ohio after May 31, 1947)
- Burlington Route and connecting lines (Note 1)
- Milwaukee Road (Note 2)
- Milwaukee Road/Union Pacific (Started October 1955)
- Pennsylvania Railroad and connecting lines (Note 3)

Note 1: Trains west of St. Paul, Minn., operated by Great Northern Ry., Northern Pacific Ry. and Spokane, Portland & Seattle Ry. The route to Kansas City shown is via the Centennial Cutoff, which opened in October 1952.
Note 2: Trains to Sault Ste. Marie operated jointly with Soo Line.
Note 3: Trains between Fort Wayne and Detroit operated by Wabash Ry. and south of Cincinnati and Louisville by Louisville & Nashville R.R. and other connecting lines.

Not to scale, not all routes shown
Based on artwork by Robert Wegner for *Trains* magazine
©2002 Kalmbach Publishing Co., used with permission

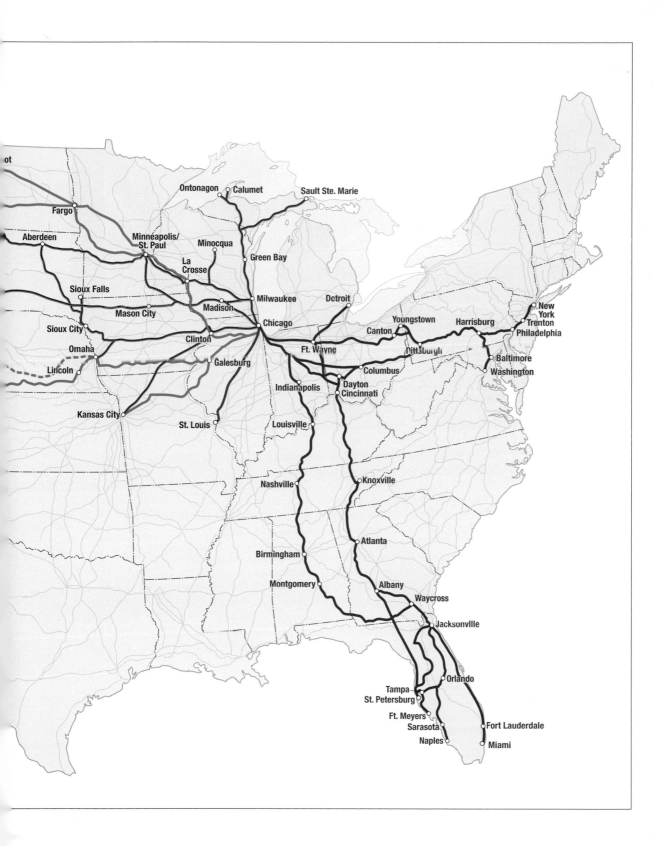

CHICAGO UNION STATION
210 S. Canal St.

Amtrak's *Floridian*, successor to the Pennsylvania Railroad's *South Wind*, waits at the south end of Union Station on June 8, 1975. One of the station's elevated baggage platforms is seen beside the train. (Verne Brummel photo)

The prevalence of competition on key routes was another hallmark of Union Station. The Burlington Route and the Milwaukee Road fiercely vied for business to Denver, Kansas City, and Minneapolis and St. Paul, MN as well as to points in the Pacific Northwest. In earlier times, the Alton Railroad also competed for business to Kansas City, which was the only city accessible from Chicago by three separate carriers from the same downtown station.

Despite the overlapping nature of these routes, Chicago Union Station—more than any other station in the city—resembled a balanced hub-and-spoke system allowing for efficient passenger connections. PRR complemented the station's western-oriented carriers and directly reached 25 of the country's 100 largest cities—the most cities served by any carrier in the city. PRR's *South Wind* served Birmingham, AL, Jacksonville, Miami, and Tampa jointly with connecting lines, while the *Detroit Limited* provided a popular jointly operated service to Detroit with the Wabash Railway.

The station, however, had shortcomings as a transportation hub. Most notably, it lacked direct service to Louisiana, New England, and the Southwest, including Arizona, southern California, and Texas. It was a weakling to most Great Lakes ports, lacking pre-Amtrak direct service to Buffalo, Toledo, and Toronto, and losing direct service to Detroit after the war. However, the gap in service to southern California was filled when the Milwaukee Road began handling Union Pacific trains east of Omaha in 1955 (these trains had previously operated over the Chicago & North Western). In the process, Union Station not only gained

direct service to Las Vegas and Los Angeles, but also had a second carrier to Denver, Portland, and the San Francisco Bay Area, making it the clear leader in transcontinental travel.

Another remarkable quality of Union Station was the longevity of its trains, particularly on the Burlington Route and Milwaukee Road. Between 1942 and 1956, the Burlington actually *increased* the number of long-distance trains serving the city, a rare occurrence during the postwar period. The strength of its postwar passenger business allowed Union Station to outdistance St. Louis Union Station as the busiest and most comprehensive passenger-train hub in the country by the mid-1960s.[18] The consolidation of former LaSalle St. trains into Union Station by Penn Central (successor to PRR) in 1968 also helped offset some of the painful cutbacks of this era. Union Station still had 64 trains in 1969—more than half the number it had in 1942, making it, on a relative scale, a success story.

Further contributing to Union Station's vitality was an expansion of the Milwaukee Road's commuter service after the war.[19] But, the station's grandeur and passenger-handling capacity were irrevocably compromised by the demolition of the passenger concourse in 1969 and its replacement with a more austere concourse, below the newly constructed Gateway Plaza office tower. The timing was inopportune, considering that almost all of Chicago's remaining long-distance trains were consolidated into the station within a few years of the creation of Amtrak in 1971. Although this consolidation gave the station an efficient system of routes stretching across the continental United States, many travelers found the boarding area unattractive and cramped.

Today, Union Station's trains directly serve 33 of the 50 largest cities in the United States, the most of any station in the country and more than those available prior to 1971. The terminal's 55 daily long-distance trains, however, pale in comparison to service levels of the thunderous wartime years. Even though an estimated 54,000 weekday passengers move through the station today—a record high (although only about one in eight are *intercity* passengers)—the historic Great Hall tends to be underutilized. As noted in Appendix I, ambitious efforts to redesign the station to improve its efficiency and enhance to the passenger experience are presently under way. ∎

CHICAGO UNION STATION
210 S. Canal St.

Several passengers walk towards an Amtrak train at a busy Union Station on July 11, 1987. Office tower construction directly above these platforms in 1969 has made the area far less inviting to passengers than before. Unfortunately, mitigating this situation has proven difficult ever since. Visible at left is a ramp leading to one of the station's baggage platforms, which are only lightly used today. (Marty Bernard photo)

Union Station's massive Beaux-Arts colonnade along Canal St.—one of its most-photographed features—provides refuge for a pigeon on June 24, 1975. (Verne Brummel photo)

A new-car display, brightly lit newsstand, and video monitors displaying information about Amtrak arrivals and departures create an interesting environment for travelers in Union Station's Great Hall on June 24, 1975. Amtrak later moved all passenger-related services, promotional displays, and retailers out of the hall, making it a somewhat desolate place. (Verne Brummel photo)

The *City of Los Angeles* hugs the Chicago River on its final departure from the north side of Union Station on April 30, 1971. This streamliner, which operated over the Milwaukee Road as far as Omaha, and over Union Pacific to points farther west, was eliminated with the start-up of Amtrak service the following day. On account of having cars destined for many different cities due to Union Pacific's consolidation of passenger trains in the 1950s and 1960s, critics often derisively referred to it as the "City of Everywhere." (George Hamlin photo)

BUS TERMINALS
Terminal Town

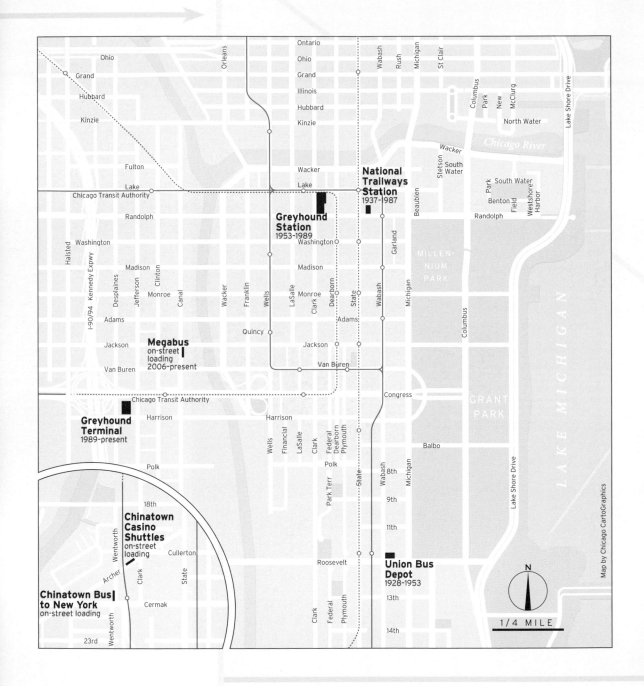

Ohio
Grand
Hubbard
Kinzie

Orleans

Ontario
Ohio
Grand
Illinois
Hubbard
Kinzie

Wabash
Rush
Michigan
St Clair

Columbus
Park
New
McClurg

North Water

Chicago River

Wacker

Fulton

Lake

Chicago Transit Authority

Wacker

Lake

**National
Trailways
Station**
1937-1987

Stetson
South
Water

Beaubien

Park
South Water

Benton
Field
Westshore
Harbor

Randolph

Randolph

**Greyhound
Station**
1953-1989

Washington

Washington

Halsted

Kennedy Expwy

I-90/94

Washington

Madison

Despiaines
Jefferson
Clinton
Monroe

Madison

Garland

Madison

Garland

MILLEN-
NIUM
PARK

Wacker
Franklin
Wells
LaSalle
Clark
Dearborn
State

Monroe

Wabash

Michigan

Canal

Adams

Adams

Columbus

Quincy

Jackson

Jackson

Jackson

Megabus
on-street
loading
2006-present

Van Buren

Van Buren

Congress

GRANT
PARK

Chicago Transit Authority

**Greyhound
Terminal**
1989-present

Harrison

Harrison

Wells
Financial
LaSalle
Clark
Federal
Dearborn
Plymouth

Balbo

Polk

Polk

Park Terr

State

Wabash
8th
Michigan

LAKE MICHIGAN

Lake Shore Drive

Lake Shore Drive

18th

**Chinatown
Casino
Shuttles**
on-street
loading

Wentworth

Archer

Clark

State

Cullerton

9th

11th

**Chinatown Bus
to New York**
on-street loading

Cermak

Roosevelt

**Union Bus
Depot**
1928-1953

13th

Clark
Federal
Plymouth

14th

23rd

N

1/4 MILE

Map by Chicago CartoGraphics

A pair of Greyhound buses waits under the suspended roof at the national operator's main station on Harrison St. In the distance, the Willis Tower (center) and 333 S. Wacker Building (right center) pierce the night sky and testify to the awkward separation between this depot and the heart of the central business district. (Gavriel Vaz photo)

T he evolution of Chicago's bus stations and departure locations reflects remarkable efficiency and flexible planning when compared to the city's railroad stations and airports. Bus stations were smaller in scale and subject to less public oversight than these other facilities. Private companies proved quite nimble in making decisions about station locations on the basis of the market's needs.

For several decades, starting in 1928, most major bus lines operated from Chicago's Union Bus Depot at Wabash and Roosevelt—a large station favorably situated near the Central and Dearborn railroad stations. Bus lines affiliated with Trailways operated from a six-story building less than a mile away at 35 W. Van Buren St.

This map shows the six principal locations in or near downtown Chicago that have been the origins and destinations of intercity bus routes since 1939. The street grid depicted is present day. Union Bus Depot at Roosevelt Rd. and Wabash Ave., and the Trailways Station at 20 E. Randolph, were dominant through the 1940s but later dwarfed by the Chicago Greyhound Station that opened in 1953. Several operators, most notably Chinatown carriers and Megabus, today use curbside locations.

Passengers departing Chicago could also step aboard buses at various "courtesy stops" made by most bus lines in the heart of the Loop and at outlying locations. Particularly large numbers of people boarded buses that stopped at 170 N. State St., where Greyhound maintained a small ticket office and picked up passengers at curbside. Passengers traveling to downtown Chicago from other cities, however, generally had to use a full-service station, because of baggage-sorting needs that could not be met efficiently at the courtesy stops.

By the early 1930s, the bus companies were clamoring for more spacious terminal buildings. Commerce in the downtown district was quickly gravitating north due to a construction boom in areas near the Chicago River and along North Michigan Ave. Both Greyhound and Trailways affiliates developed plans to relocate their terminals to the Loop proper, with the latter moving first by settling in to a new "ultra-modern" National Trailways Station at State and Randolph. Greyhound, after several false starts, finally opened a station at Clark and Randolph in 1953.

Among these, the Greyhound Station stands out, being one of only two new large-scale intercity transportation terminals built in the metropolitan Chicago region after World War II (the other was O'Hare Airport). Equipped with 31 bus bays (stalls for loading and unloading buses) and more than a dozen retail business, it was likened to a city within a city. Passengers could depart for many destinations at almost all hours of the day, and retail businesses were open around the clock to accommodate them. Both this terminal and the nearby Trailways depot also afforded patrons excellent access to the city's rapid-transit system—something most of Chicago's downtown rail stations lacked.

The public image and viability of scheduled bus service took a sharp turn for the worse in the late 1960s. More-affluent passengers turned their backs on motor-coach trips in favor of airplanes and private automobiles, while reports of escalating crime at stations made bus travel less desirable than other alternatives. Between 1960 and 1990, the number of intercity bus arrivals and departures dropped from 454 to 290, and developers began eyeing the Greyhound station's prime real estate for a modern office building. The Trailways station closed in 1987.

When Greyhound and other bus lines moved to the much smaller Express Terminal at Harrison and Clinton in 1989, the intercity bus industry surrendered one of its key advantages: having a station location in the heart of downtown. Since then, most intercity bus lines have been relegated to this peripheral location. The move to the new station improved opportunities for transfers between intercity buses and trains departing from Union Station, which was only four blocks away, but this convenience remains unfamiliar to most travelers to this day. Chicago never achieved the synergy between intercity rail and buses enjoyed by Boston, New York, and Washington, D.C., where bus and rail stations enjoyed greater proximity.

Signs of a recovery in intercity bus travel emerged in the late 1990s as low-cost operators, responding at first to demand from Asian-American communities, began linking various urban Chinatown districts along the East Coast, with one route stretching all the way to Chicago. It was not until 2006, however, that a true renaissance took place. Larger numbers of passengers, particularly young people, flocked to Megabus, which established a hub on the 200 block of South Canal St. next to Union Station. Passengers were attracted by the company's bargain fares, express service, and double-decker buses. As in most other cities of operation, Megabus served its local clientele without a station, departing instead from a curbside stop.

Transportation planners eager to integrate buses and trains more effectively into a hub-and-spoke system by consolidating services into a single terminal will likely have to bide their time until new opportunities present themselves. As noted in this book's conclusion, however, the remarkable growth of this sector makes it inevitable that changes are around the corner. ■

(Opposite) A Greyhound bus painted in the Trailways livery heads down the Stevenson Expressway on July 7, 1989. Greyhound had absorbed various Trailways affiliates earlier that year, but the repainting of its equipment was not yet complete. (Ron Sullivan photo, John LeBeau collection)

UNION BUS DEPOT
1157 S. Wabash Ave.

Union Bus Depot's façade, clad in terra-cotta, survived virtually unchanged in 2013—60 years after it was abandoned by intercity bus companies. Along with Dearborn Station's head house and Meigs Field's Main Terminal Building, it is one of only a handful of large terminals in Chicago that have been successfully repurposed. (Xhoana Ahmeti photo)

Union Bus Depot, a large two-story structure on the northeast corner of Roosevelt Rd. and Wabash Ave., was Chicago's principal intercity bus station for more than 25 years. Opened in 1928, the station handled more than 300 buses per day well into the postwar era, making it the Midwest's busiest bus terminal. Designed by architect David Saul Klafter, a protégé of Daniel Burnham, this graceful building is distinctive for its elegant white terra-cotta exterior and ornate windows. Greyhound Lines briefly maintained its company headquarters on the second floor.

The station served all lines affiliated with Greyhound, including the widely used Overland Greyhound lines, as well as several independent lines, such as Indiana Motor Bus and Indian Trails. Buses loaded both in front and in the rear of the rectangular building. Although the terminal was considered inadequate by the late 1930s, Greyhound's effort to build a replacement stalled, resulting in considerable overcrowding and operational inefficiencies. In 1943, there was discussion about installing a helicopter pad on the roof to support a new aerial service being contemplated by Greyhound—a futuristic operation that never came to be.

Union Bus Depot's location at 1157 S. Wabash might seem too far south for a centralized bus station today. At the time it was built, however, it was highly advantageous, being catercorner to a site envisioned for a great railroad station in Daniel Burnham and Edward Bennett's 1909 *Plan of Chicago.* Although that station was never built, passengers benefited from the bus depot's proximity to several of the existing downtown railroad stations, particularly Dearborn St., which was only four blocks away, and Central Station, a block and a half away. Streetcars ran immediately outside Union Bus Depot, and the North Shore Line's electric trains, and elevated rapid-transit lines stopped just a half block west at Roosevelt Rd.

By the late 1940s, approximately 300 weekday bus trips used the station, with the number rising to 500 on holiday weekends. Thousands of African Americans migrating from the Deep South to the industrial Midwest passed through its doors, attracted by the fact that bus service was generally less expensive than rail service, albeit appreciably slower due to roadway congestion and the number of stops (including rest stops) generally made by buses. The fastest bus trip from Memphis in 1950, for example, took 16 hours, compared to 8 hours and 12 minutes on the fastest Illinois Central streamlined train.

Greyhound Lines timetable, effective May 28, 1934. (Author's collection)

A woman talks to a passenger departing on a streamlined Greyhound Lines "Silverside" bus at Union Bus Depot in September 1943. This facility's constricted loading area, evident in this photo, was woefully inadequate by the late 1930s as both passenger counts and vehicle sizes increased. Efforts to replace the terminal, however, remained on hold. (Esther Bubley photo, Library of Congress)

UNION BUS DEPOT
1157 S. Wabash Ave.

A Chicago Surface Lines streetcar built by Brill in 1910 rolls south down Wabash Ave. in front of Union Bus Depot on September 27, 1947. Four days later, CSL operations were taken over by the newly established Chicago Transit Authority. A distinctive Greyhound "Silverside" bus can be seen directly in front of the terminal's Roosevelt Rd. entrance, which has a canopy adorned by the company logo (far right). Greyhound had its corporate headquarters on the second floor. (Krambles-Peterson collection)

Most departures from Union Bus Depot made curbside courtesy stops in the Loop to pick up passengers. Greyhound's stop at 170 N. State St. was particularly important in this regard. Travelers headed east or south, meanwhile, could board in the city's South Side Woodlawn neighborhood. Courtesy stops were not feasible on inbound runs because checked baggage was too cumbersome to unload at multiple curbside locations. Union Bus Depot, consequently, saw far more arriving than departing passengers.

Unlike all other historical bus stations featured in this volume, this depot remained at full capacity until its end. Even so, the number of bus companies making this their principal station gradually diminished, falling from nine in 1943 to seven in 1953, largely due to the consolidation of many smaller operators into the Greyhound system. Service ended abruptly on March 19, 1953, when Greyhound opened its new station at 170 W. Randolph St. Since then, the older building has been used for many different functions. It served as a union hall and small grocery store before being rendered vacant. Today it is a Trader Joe's supermarket and Potbelly Sandwich Works outlet. The building's façade, however, is virtually unchanged since the final motor coach departed more than 60 years ago. ■

At least two servicemen are in the queue that has formed for an outbound Greyhound bus at Union Bus Depot in 1942. The extensive amount of luggage beside the passengers suggests the destination may be many miles away. This is one of only a few interior photos of the station known to exist. (Esther Bubley photo, Library of Congress)

The former Union Bus Depot, viewed from the south side of Roosevelt Road in November 2013, remains a South Loop landmark. The station's proximity to the Roosevelt "L" station (left center) made it convenient for passengers arriving on rapid transit and the North Shore Line. (Author's collection)

National Trailways Station was aptly described by the *Chicago Daily News* as the city's "first modern loop union bus terminal."[20] This attractive Art Deco facility—the city's second-busiest bus station throughout its entire 54-year existence—is remembered for its streamlined appearance and blinking neon lights. Passengers could board buses to an enormous range of destinations, albeit with fewer frequencies than were available at the rival Greyhound station several blocks west on Randolph St.

In the midst of the Great Depression, on February 27, 1937, 13 bus lines, including all the affiliates of Trailways serving Chicago, moved their operations out of an older facility at 35 W. Van Buren into this glistening facility inside the Loop. Developed by the Santa Fe Trail bus system and Burlington Trailways, the new station at 20 E. Randolph St. was designed by the famous architectural firm of Graham, Anderson, Probst & White (which also designed Chicago Union Station). Reports had circulated that Greyhound and other major carriers operating from Union Bus Depot would use the new station.[21] Most carriers at this rival station, including Greyhound, however, opted to stay put.

Travelers could pass the time in air-conditioned comfort in the waiting room and in an around-the-clock lunch room. The station also had its own barber shop and newsstand, and National Trailways Corp., the umbrella organization to which all Trailways operators belonged, had offices on the second floor. Motor coaches arrived and departed from diagonal bus bays on the station's west side.

Rising traffic put a heavy burden on the station. An estimated 135 buses used the terminal daily at its opening, but the number of passengers surged

(Opposite) Late on a May 1986 night, the Chicago Trailways Station's brilliant neon marquee illuminates Randolph St. Mammy's Restaurant, the popular eatery inside the station, however, is closed. This photo was taken around the time of the 50-year anniversary of the station's opening. Three years later, Greyhound bought its rival Trailways and closed the facility. (Mel Bernero collection)

The National Trailways Station—considered one of the most modern bus stations in the country—was decorated in patriotic fashion at its opening in early 1936. This "ultramodern" station, shown here facing north across Randolph St., had bus bays on its west side. Note the changes in the signage in this grand-opening photo and the photo on the previous page. (Author's collection).

NATIONAL TRAILWAYS STATION
20 E. Randolph St.

A Continental Trailways GMC PD-4106 bus is parked between trips at its assigned bay at the Chicago Trailways Station, circa 1965. This bus would likely have exited via the alley at the north end of the terminal to reach State St. or Wabash Ave. (Mel Bernero collection)

during World War II. Traffic remained strong after the war. The station had 96 intercity departures and several dozen commuter runs in 1946, and the number of companies using the station (11) remained steady for many years. Burlington Trailways and its American Bus Lines affiliate, which together operated 29 daily bus trips, had departures to points as far away as the West Coast.

Trailways and Greyhound (which moved into a massive new terminal at 170 W. Randolph St. in 1953) maintained a fierce rivalry, and their neon signs were familiar complements to the flashing marquee signs of the theaters lining Randolph St. In some respects, the status of the Trailways depot was akin to Midway Airport today, while the Greyhound station was comparable to O'Hare, in traffic and recognition. Each bus terminal had expansive route networks, but the Trailways operation was only about a third as large. Unlike the city's airports, which tend to be distinct hubs, the ease of transferring between the bus stations created synergy that made the whole system greater than the sum of the parts.

The Trailways station's services diminished at a slower pace than those of the Greyhound facility as the industry fell on hard times. During the 1950s and 1960s, several of the original Trailways affiliates, including Burlington, were consolidated into Continental Trailways, which maintained its primary hub in Chicago. Even so, the number of daily buses dropped to just 30 in 1980, rendering the terminal significantly oversized. The facility became redundant after Greyhound acquired Continental Trailways on July 14, 1987. Barely a month later, on August 19, all buses, including those with no affiliation to Continental Trailways, were transferred to the Greyhound station.

The Trailways station was demolished a few years later to make room for a public parking garage and retail development, but the former driveway on the west side of the building survives as an alley. ■

Trailways Chicago-area timetable, effective November 1, 1951.

A Continental Trailways bus is in rival territory, directly in front of the Chicago Greyhound Station at Clark and Washington streets. The Golden Eagle model coach in this August 1964 photo has likely just departed the Trailways station at 20 E. Randolph. The Sherman Hotel is at left, Marina City is in the distance. (John LeBeau collection, courtesy of Melvin Bernero)

NATIONAL TRAILWAYS STATION
20 E. Randolph St.

This rare photograph shows the side of the Trailways Station's aging Art Deco façade circa 1965. A Continental Trailways "Thru Liner" (model GMC PD-4104) is in the foreground, while the older coach behind it appears to be from the Cardinal Bus Company. Cardinal offered commuter service between the Loop and north-suburban Evanston. (Mel Bernero)

A Continental Trailways "Eagle" has just passed under the Wabash Ave. "L" and is approaching the Trailways Station on Randolph St., circa 1973. This photo was taken from near the station's front door. Marshall Field's decorations suggest that this was during the holiday season. (Ron Sullivan photo, John LeBeau collection)

The oversized marquee of the Trailways Station, with the Mammy's Restaurant sign directly below it, made these Randolph St. institutions difficult to miss. This photo was taken on June 19, 1987 for a newspaper report about a possible Greyhound–Continental Trailways merger. Both the station and its restaurant, however, would be gone before the decade ended. (Maureen Collins photo, *Chicago Sun-Times*)

CHICAGO GREYHOUND STATION
170 W. Randolph St.

Spanning an entire block of N. Clark St. between Randolph and Lake, the Chicago Greyhound Terminal was a beehive of activity for decades. The massive facility compensated for its lack of exterior charm by having an innovative design that allowed arriving and departing buses to reach the station below street level. In this photograph, circa 1980, the Clark and Lake "L" station is visible in the distance. (Paul Dimler photo)

The Chicago Greyhound Station at the corner of Randolph and Clark streets was the largest independently owned bus station in the world for many years. Home to more than a dozen retailers and a waiting area 280 feet long, it was the most important mid-American transfer point for intercity bus passengers throughout its entire 47-year life. Buses arrived at and departed from underground bus bays using passages and ramps linked to Lower Wacker Dr. This arrangement contrasted with the more congested street-level access of the competing National Trailways Station.

The five-story Greyhound station, opened on March 19, 1953, had three floors above the street and two levels below it. On the ground floor, travelers enjoyed the convenience of 13 businesses with doors opening to the station's arcade and the street. The *Chicago Daily News* described it as the "Little City of Shops."[22] The two upper floors were used for office space (much of it occupied by Greyhound's headquarters), while the two underground levels were used for a waiting room, ticketing area, and bus arrivals and departures. The rooftop had several hundred parking spaces accessible by ramps.

The effort to build this station spanned almost two decades. Having outgrown the cramped and congested Union Bus Depot at Roosevelt Rd. and Wabash Ave., Greyhound purchased property in the heart of the Loop for a new terminal in 1941. However, the company's proposal to build a sleek, modern structure designed by W.S. Arrasmith, the architect of many of Greyhound's other big-city depots, stalled. The country's deepening involvement in World War II made such a large-scale project a practical impossibility.

After the war, Greyhound abandoned the Arrasmith plan and hired the architecture firm of Skidmore, Owings & Merrill to design a replacement terminal. Skidmore envisioned an elaborate station with a 225-foot "private tunnel" linking the station to Garvey Ct., an underground street connected to Lower Wacker Dr. Creating the tunnel required razing a four-story building on Lake St. Fresh air needed to be pumped into the tunnels and underground levels using a sophisticated forced-air ventilation system. The company considered whether the benefits were worth the expense, but hedged its bets by designing the station to support additional office development. The structure was designed to accommodate the eventual construction of a 20 story tower, echoing the planned (but unbuilt) office structure above the Great Hall at Chicago Union Station.

Mayor Martin H. Kennelly and Governor William G. Stratton were on hand to commemorate the opening of this enormous $10 million terminal, which, according to the *Daily News*, was the first multimillion-dollar building to be

The Greyhound Station's passenger concourse and waiting room, one level below the street and one level above the loading concourse, boasted an around-the-clock ticket counter, restaurant, and classic circular information kiosk (at center). The upper floor, the "Little City of Shops," offered abundant retail options. The ticketing windows were to the left, behind the escalators. (Author's collection)

completed in the Loop in nearly 20 years.[23] Although lacking the architectural detail of the earlier Arrasmith proposal, the *Chicago Tribune* hailed it as one of the most modern stations in the world. The waiting room, situated one floor below the street, featured several retailers, 800 storage lockers, and a ceiling two stories tall in certain places. When buses were ready for boarding, passengers descended to the glass-enclosed loading concourse one floor below the waiting room (and two floors below the street).

Approximately 300 daily buses used the terminal upon its opening, with up to 200 more operating during busy holiday periods. The station had a capacity for 124 buses per hour, based on four hourly departures from each of the 31 bays. To support the heavy flow of traffic, Greyhound required several of the businesses in the station to remain open at all hours. The number of daily arrivals and departures dropped only modestly through 1960, and the expansion of the interstate highway system greatly improved the speed of service to Boston, Miami, San Francisco, and other distant cities.

As the intercity bus gradually lost its luster, passenger loads dropped sharply in the 1970s. By 1980, the number of daily buses had fallen to around 250, and reports of crime blemished the station's reputation. The decline in service accelerated as middle-income passengers opted for other modes of transportation, a process only temporarily interrupted when buses previously using the National Trailways Station at 20 E. Randolph St. began using this station in 1987. By the late 1980s, Greyhound was eyeing property to build a

(Opposite) The new State of Illinois Center (now the James R. Thompson Center) nears completion, towering above the Chicago Greyhound Bus Station in 1983. The construction of the center portended a sharp rise in demand for downtown office space, which would eventually doom the station. The pedestrians in the foreground are on Daley Plaza. (C. William Brubaker photo)

The glass-enclosed loading concourse, two stories below street level, linked to the main passenger concourse by stairway and escalator, had 26 bus bays. Clustering the gates in this relatively small area allowed for easy bus connections. As is evident by the sign, departures were typically assigned to one side of the concourse and arrivals to the other. (Chicago History Museum; William C. Hedrich photo; HB-16086-H)

Shown here from a vantage point east of the Randolph St. passenger entrance, circa 1965, the proximity of the Greyhound Station to other prominent businesses, including the Sherman Hotel and Woods Theatre, helped sustain the many retailers inside the terminal. The Toffenetti restaurant and cocktail lounge (center), was one of more than a dozen businesses with entrances both inside the terminal and on the street. *Call Me Bwana*, a comedy featuring Bob Hope, was playing at the theater. Louis Sullivan's famous Garrick Theater building was sadly razed to make room for the nondescript parking garage between the theater and station (Charles W. Cushman Collection, Indiana University Archives, P12892)

CHICAGO GREYHOUND STATION
170 W. Randolph St.

much smaller terminal, possibly one integrated with a new office complex, and it sold the land occupied by its existing station for real estate development.[24] Ambitious plans by the city government to redevelop the North Loop, which called for cleaning up the area's image and transforming Randolph St. into a theater district, raised questions about the station's future.

Even as its image worsened, however, the Chicago Greyhound Station remained a prominent downtown landmark, attracting many consumers who were not traveling by bus. The station's first floor boasted some of the city's most familiar retailers, including Ronny's Steak House, Order by Horder, and Sam the Shoe Doctor. The Greyhound Post House restaurant and the Hide-a-Way cocktail lounge were also popular with travelers and non-travelers alike.

Such retailing aside, the station's best days were clearly in the past, and on December 6, 1989, the roughly 150 daily scheduled bus operations that remained were transferred to the new Chicago Greyhound Terminal at 630 W. Harrison St. *Chicago Tribune* reporter Robert Davis summed up the move by writing, "Grand openings often are accompanied by not-so-grand closings, and so it was in Chicago's Loop Wednesday, when the Greyhound Bus Station at Randolph and Clark streets unceremoniously closed its doors, ending a 36-year era as an urban Ellis Island for travelers."[25] The city tore down the old station to make room for the Chicago Title & Trust Building, now one of Chicago's most distinctive skyscrapers.

A Greyhound Americruser 2 is at Bay 17 on December 2, 1989, just a few days before the massive Chicago Greyhound Station was closed. The escalator linking the loading zone—two floors below the street—to the main waiting room can be seen through the glass window at right. (Ron Sullivan, John LeBeau collection)

Few preservationists expressed regret over the station building's demise, due to its rather austere exterior. With the revival of bus travel in recent years, however, the advantage of having such a station in the heart of the Loop has risen. One can only speculate how the massive Greyhound station—and its private tunnels—would be used today had they survived. ■

A Greyhound bus has emerged from uncongested Lower Wacker Dr. and is approaching the Upper Wacker and Lake St. intersection on July 3, 1989. In the distance, the Merchandise Mart built on the former site of the original Chicago & North Western Station, and the Helene Curtis Building (upper right) can be seen. This bus may be heading back to the storage yard. (Ron Sullivan, John LeBeau collection)

The end is near for the Chicago Greyhound Station on December 2, 1989. The ticketing area has become too large for the facility's remaining business. The sign at left reads "Don't Miss the Move" and alerts patrons of the transfer of station operations to 630 W. Harrison St. (Ron Sullivan, John LeBeau collection)

This aerial photograph of the Chicago Greyhound Station was taken on February 26, 1953, only a few weeks before its grand opening. The terminal's ground floor was being readied for retail use and the two windowless upper floors were being prepared to serve as Greyhound's company offices. The company envisioned the rooftop parking one day giving way to an office tower. Clark St., still equipped with streetcar lines, is at left. (Bob Kotalik photo, author's collection).

CHICAGO GREYHOUND TERMINAL
630 W. Harrison St.

The Chicago Greyhound Terminal at 630 W. Harrison features a dramatic cable-suspended roof. This station's location near the Kennedy Expressway limits the risk of delays due to street congestion, but at the expense of convenience for many passengers traveling on foot. The facility (also called Chicago Greyhound Station) is five blocks from Union Station and more than six blocks away from the Loop district. (Author's photo)

The Chicago Greyhound Terminal, owned by Greyhound Lines, is the busiest conventional intercity bus terminal in the central United States. Bounded by Harrison, Des Plaines, and Jefferson streets, this new station replaced the much larger Greyhound station at Clark and Randolph streets in 1989.

As the intercity bus industry fell on hard times in the 1980s, Greyhound began looking for a site to build a much smaller station. The carrier initially considered a site near the Addison St. interchange on the Kennedy Expressway, an idea strongly opposed by this middle-class—and predominately white—neighborhood, partially out of concern over the image of intercity bus travel as a mode of travel mostly for the poor. Greyhound later purchased a site near the famed Circle Interchange where the Congress, Dan Ryan, Eisenhower, and Kennedy expressways meet.[26] This location eliminated the need for buses to travel to and from the middle of the Loop, with the potential to greatly reduce operating costs and shave several minutes of travel time from each trip.

Architects Nagle, Hartray & Associates designed the station to have a cable-suspended roof and a brick façade along Harrison St. The suspended roof obviated the need for columns inside the terminal and created a floor plan conducive to efforts to reduce crime. A skylight atrium several stories high allowed natural light to reach the waiting room. The new station had 24 bus bays (compared with 31 at the old facility) divided equally between the east and west sides of a large waiting room. Dining was limited to a cafeteria, leaving customers without the choices they enjoyed at the old Greyhound station.

Music from a University of Illinois at Chicago band enlivened the ribbon-cutting ceremony on December 7, 1989. In the first month after its opening, the terminal served 158 weekday buses of five carriers, with Greyhound Lines accounting for about three-quarters of the total. Southeast Trailways, the second largest carrier, had just 16. Like its predecessor, the station boasted a relatively even mix of services to the east, west, north, and south of the city, making it Greyhound's most important mid-American hub. Many passengers transferred between this terminal and Chicago Union Station, four blocks away. Others used rapid-transit service from the Clinton St. subway stop, just two blocks away.

The next decade, however, was a difficult one for bus operators and riders alike. Greyhound's owners filed for bankruptcy protection in 2001 and continued to trim services, with cuts beginning in 2004 and lasting through

The ticketing area and information counter at the Chicago Greyhound Terminal, considered a showpiece when opened in 1989, is decorated in the carrier's familiar blue-and-white colors. These facilities are less than half the size of those of the older station at 170 W. Randolph. (Author's photo)

CHICAGO GREYHOUND TERMINAL
630 W. Harrison St.

The waiting room at Greyhound Express Station, like that of the former Trailways Station, offers direct sightlines to all of the depot's 24 bus bays. The suspended roof eliminates the need for columns inside the terminal and made security easier to provide. (Author's photo)

late 2005. By the end of that year, the number of weekday arrivals and departures declined to just 125, and all through-bus service to Florida and the West Coast ended, as Greyhound trimmed its system to focus on short- and middle-distance routes. For the first time in a half century, bus travelers could not directly reach major cities in these regions or travel cross-country with a simple transfer in Chicago.

A modest turnaround began after Greyhound completed a $60 million nationwide overhaul of its stations in late 2007. Chicago Greyhound Terminal received plasma screen televisions and new signage, and renovated bathrooms. In 2010, Greyhound introduced a premium "Express" service featuring onboard Wi-Fi, power outlets, and guaranteed seating. To accommodate both Express and conventional traffic on the same buses, the company began the unusual practice of having some buses make two stops in the station—first on a west bay to pick up Express passengers (thereby assuring these passengers a seat) and another a few minutes later at an east bay to pick up conventional ticket holders.

Despite recent expansions to Greyhound Express, the terminal has not yet regained the level of high-frequency service once so common on Chicago's intercity bus network. The station today has just 108 weekday arrivals and departures—all but 23 operated by Greyhound, and only about a third of 1980 traffic levels. Although Greyhound never reinstated direct service to Florida or the West Coast, passengers can still reach more destinations from Chicago Express than from any other Midwestern bus station.

This station's peripheral location has had lasting consequences. Even as downtown Chicago's economy expands, few consider the station as having the potential to assume a transportation role even remotely resembling that of the Port Authority Bus Terminal in New York, which today serves a vibrant mix of intercity and commuter buses. If this terminal had been situated closer to the urban core (the nearest station on the Loop elevated system is eight blocks away), it might have played a greater role in the recent revival of regional and long-distance bus service. (Most newly created bus routes serving Chicago use curbside locations closer to Union Station). Another problem resulting from the Harrison St. location is the perception of inadequate safety. Although the location is relatively free of congestion, it is not considered a desirable place to wait for extended periods because of concerns over crime. ■

A brightly painted Burlington Trailways bus and a pair of Greyhounds are at the west bay at the Chicago Greyhound Station in September 2013. The frequent service provided by the regional Trailways operator to Illinois' Quad Cities as well as points in Iowa and Nebraska made it second in size only to the station's owner. (Xhoana Ahmeti photo)

CHICAGO GREYHOUND TERMINAL
630 W. Harrison St.

The new Chicago Greyhound Station, identifiable by its cable-suspended roof, is busy on this cloudy afternoon in 2006. A pair of coaches can be seen at the bus bays, while several others are visible on the apron. In the distance, beyond the enormous Central Post Office, is the viaduct carrying LaSalle Street Station's tracks across Harrison St. (far right). One Financial Place (the slender brown building at center) was built on the site of that station's former head house. Buses headed for the Greyhound Station exit the Congress Expressway (left) at a ramp just west of the post office. (Steven Vance photo)

CHINATOWN BUS STOP
2130 S. Wentworth Ave.

A Three Happiness Bus Lines coach waits on Wentworth Ave. within view of the platforms of the Cermak-Chinatown Red Line Stop in 2012. Like equipment used by many Asian operators, the coach is devoid of company lettering or insignia. (J. Dalide photo)

The curb at 2130 S. Wentworth Ave., adjacent to CTA's Cermak–Chinatown Red Line Station, serves as a staging area for a daily bus service to New York City and shuttle service to Chicago–area casinos. This heavily used loading zone remains devoid of any signage indicating its role as an intercity bus stop.

By the early 2000s, Asian-American communities across the United States were embracing low-cost bus service, resulting in a proliferation of "Chinatown Bus" service all over the country. Although this sector was initially confined to major cities on both coasts, its expansion into the Midwest was perhaps inevitable due to Chicago's size and the dynamism of its Chinatown district. The metropolitan region has nearly 70,000 people of Chinese ethnicity, and Chinatown boasts the greatest concentration of businesses, institutions, services, and housing catering to this growing ethnic community.

In addition, the Chinatown feeder ramp, a branch off the Dan Ryan Expressway, made this neighborhood a convenient drop-off point for all types of bus operators seeking to avoid traffic congestion in the urban core. Contrary to what designers of the feeder had anticipated before its completion in 1962, the road was for many years underutilized. Recently, however, it has become a popular way to avoid traffic backups near the Circle Interchange west of the Loop business district.

The Chinatown service to New York City appears to have begun in the early 2000s and has never been more than a single daily arrival and departure. In late 2013, a Three Happiness Bus Lines bus—generally a white coach devoid of company insignia—was departing on some days at 7:30 p.m. and arriving at the curb at 156 E. Broadway in New York's Chinatown neighborhood at 10 a.m. the following day. The westbound trip leaves New York at 9 p.m. and arrives in Chicago at 10:30 a.m. the following day. This 789-mile route is the longest operated by a bus line serving Chicago without an intermediate stop (except for obligatory rest stops as mandated by law). A typical fare is around $80.

Three operators serving casinos in East Chicago, Gary, and Hammond, IN, meanwhile, use a location adjacent to the Red Line station on Archer Ave., just east of Wentworth. The surrounding community relies heavily on shuttle buses to reach such entertainment destinations, and they account for the majority of passengers on many departures. Hammond's Horseshoe Casino's operator offers eight weekday roundtrips, making it the most frequent casino service at this location. ■

The daily coach to New York, shown at the curb at 2130 S. Wentworth, is Chicago's only "Chinatown bus" departure. This overnight run is also the longest scheduled bus trip in the Midwest, with no intermediate stops to pick up or drop off passengers. Passengers, however, are afforded several rest stops. (John Dalide photo)

MEGABUS BUS STOP
300 Block of S. Canal St.

Passengers prepare to board a double-decker Megabus bound for St. Louis from the S. Canal St. curb in 2013. While lacking a traditional station, the local operation of this Coach USA subsidiary has grown to encompass 80 weekday departures. (Xhoana Ahmeti photo)

The stretch of Canal St. between Jackson and Van Buren streets, one block south of Union Station, is one of the largest curbside pickup and drop off points for intercity buses in the United States. There is no physical station, so passengers wait at the curb for their buses.

Megabus is part of a new genre of bus operators that eschews traditional stations in favor of curbside pickup. The carrier provides Wi-Fi, onboard power outlets, and other previously unavailable amenities at steeply discounted prices. Ticketing is handled almost entirely through the Internet. The Megabus operation at this location has grown to more than 80 daily bus trips.

This subsidiary of Coach USA (an entity owned by Stagecoach Ltd., a British company) opened its Chicago hub on April 10, 2006. Service initially consisted of 32 buses (16 round trips) each weekday between eight Midwestern cities: Chicago, Cincinnati, Columbus, Detroit, Indianapolis, Milwaukee, Minneapolis, and St. Louis.[27] The following year, service expanded to 42 daily bus operations, with added service to Kansas City. Buses to Memphis (at 533 miles, the company's longest route from Chicago) began in 2008, while a route to Des Moines was added in 2010 and extended to Omaha the following year. Louisville service started in 2012.

Until 2010, Megabus operated from 225 S. Canal St., directly in front of one of the main entrances to Chicago Union Station. After the city council passed an ordinance in June of that year limiting use of this location to bus lines primarily carrying passengers making connections to and from trains, Megabus moved one block south to its present location, easing concerns over congestion caused by bus operations. Several other bus lines, including Van Galder (another Coach USA affiliate) and Amtrak Thruway buses, continue to operate from the "front door" of the railroad station.

Megabus now serves 20 cities from Chicago and runs buses southeast to Nashville, west to Omaha, north to Minneapolis, and east to Cleveland. Fares typically undercut Amtrak by 25 percent or more. In 2012, Megabus began selling tickets that involve bus-to-bus connections on trips through Chicago, for which schedules are sufficiently cushioned to accommodate late arrivals.

Discount operators, such as Megabus and its rival BoltBus, have similar operations in other cities, including Boston, New York, Philadelphia, and Washington, D.C. Drop-offs and pickups in several of these cities, however, have been moved to off-street locations, and in some cases, to traditional bus stations. Megabus is currently in discussions with city and state transportation officials about moving its Chicago operations to an off-street parcel—presently a parking lot—under the Congress Parkway viaduct just west of Clinton St. ■

A pair of Megabus coaches are at the carrier's busy Chicago boarding location. Passengers peering over the sidewalk wall can see the canopies above the tracks at Union Station's south concourse. Riverside Plaza is in the distance. (Xhoana Ahmeti photo)

The Megabus arrival and departure area on Canal St., only steps from the Chicago Union Station headhouse on September 5, 2013. Those waiting at the far right can look down on the trains at the station's south concourse. City planners have envisioned the possibility of moving loading and unloading to an off-street facility nearby to ease congestion. (Xhoana Ahmeti photo)

ELECTRIC INTERURBAN RAILWAY TERMINALS
Terminal Town

Ohio
Grand
Illinois
Hubbard
Kinzie

Rush

Chicago River

Orleans

Fulton
Chicago Transit Authority

Wacker
Lake

Randolph

Washington

Randolph Street Station
South Shore Line

Halsted
Union
Desplaines
Jefferson
Clinton
Canal
Market
Franklin
Wells
LaSalle
Clark
Dearborn
State
Wabash
Garland
Michigan

Madison
Monroe

Wells Street Terminal
Chicago, Aurora & Elgin

Adams

Quincy

Jackson

Adams/ Wabash Station
North Shore Line

Columbus

Lake Shore Drive

Chicago Transit Authority
Chicago, Aurora & Elgin

Van Buren Street Station
South Shore Line

Van Buren

Congress Baggage Terminal
North Shore Line

Congress

GRANT PARK

L A K E   M I C H I G A N

Harrison

Sherman
LaSalle

Balbo

Polk

Clark

State

Chicago Transit Authority
North Shore Line

Wabash

Michigan

8th

Taylor

9th

11th

Roosevelt Road Station
South Shore Line

Roosevelt

Roosevelt

Roosevelt Road Station
North Shore Line

13th

Maxwell
Halsted

14th

N

1/4 MILE

Map by Chicago CartoGraphics

A South Shore Line train leaves Randolph St. Station on a winter day in 1976. The skyline behind the train (note the iconic Carbide & Carbon Building, with its spire resembling the top of a champagne bottle) attests to the density of development near the station, which cultivated the commuter business and helped the carrier fend off abandonment before a government agency acquired it. (Mark Llanuza photo)

Interurban railways were a familiar part of Chicago's metropolitan transportation scene well into the post-World War II era. These "traction" companies operated trains that drew their power from an overhead electrical wire ("catenary") or, less frequently, from an electrified third rail near ground level. Unlike streetcar companies, which operated almost solely on city streets, interurban railroads often used private right-of-way in suburban and rural areas, allowing them to operate at relatively high speeds. Finding a way to bring trains to the heart of the city, however, was often a problem, and only a few such operators managed to reach terminals located downtown.

By the mid-1930s, the region's interurban railroad system had dramatically thinned, and only three companies—the Chicago, Aurora & Elgin; the Chicago South Shore & South Bend; and the Chicago, North Shore & Milwaukee—survived. All trains into downtown Chicago, used either the tracks of rapid-transit companies or the steam railroads. Unlike many of their steam-railroad counterparts, which tended to have just one downtown station, all three interurban railroads serving Chicago could offer customers the conveniences associated with multiple stops near the heart of the city. Still, each had one main downtown station, and these are discussed on the following pages.

The main stations of these carriers—North Shore Line Station at 223 S. Wabash Ave., Randolph Street Station at 151 E. Randolph St. (CSS&SB), and Wells Street Station at 314 S. Wells St. (CA&E)—each had ticket counters and waiting rooms that allowed passengers to reach all platforms without being exposed to the elements. Each played an important, somewhat unappreciated, role in the city's postwar transportation history. Among these terminals, only Randolph Street Station, which has been completely rebuilt, remains in use by an electric interurban railway. ■

(Opposite) This map shows the stations served by electric interurban railways in downtown Chicago as of January 1, 1950. Randolph Street Station was the terminus for the South Shore Line, while the Wells Street Station was the Chicago, Aurora & Elgin Railroad's endpoint. The North Shore Line, operating over the Loop Elevated, had its main station and ticket office at Adams and Wabash, but used the nearby Roosevelt Road Station as its terminus at the time.

NORTH SHORE LINE ADAMS/WABASH STATION
223 S. Wabash Ave.

A northbound Electroner glides into a crowded Adams/Wabash Station circa 1960. Southbound trains did not use this stop, traveling instead via Wells and Van Buren streets before turning south at the junction at Wabash and Van Buren, visible in the distance behind the train. (Gerald Widemark photo)

For more than 40 years, the 200 block of South Wabash Ave. was home to the main station and ticket office of the Chicago, North Shore & Milwaukee Railroad (North Shore Line), an electric interurban company linking Illinois' and Wisconsin's largest cities. The North Shore Line operated its trains through downtown and along the north lakefront to Evanston over CTA's elevated tracks (the "L") and over its own right-of-way beyond the city limits. This storefront station, located next to the Adams/Wabash "L" stop, had a ticket office, waiting room, lunch counter, and other passenger conveniences.

The 200 block of South Wabash became important to intercity travel in 1919, when the North Shore Line began using the Northwestern Elevated (part of which is today's CTA Red Line) between Howard St. and the northern edge of the Loop. The company transformed a storefront at 209 S. Wabash Ave. into a station. Passengers boarded trains at the Adams–Wabash "L" stop located immediately west of the station. In a somewhat unusual arrangement, however, southbound runs generally did not stop at the station, operating instead via the Loop's Wells and Van Buren streets before turning south toward Roosevelt Rd. Travelers making a round-trip from the Loop, consequently, had to return to a different Loop station from the one they departed. (There were exceptions, such as the carrier's special trains that made a "trip and a half" around the Loop to take shoppers to Marshall Field's and other department stores).

Although the Adams–Wabash stop was listed in timetables as the North Shore's "main station," it was never a point of origin for trains or the only downtown stop. Trains originated at the Roosevelt "L" station and then, starting in 1922, those with parlor and dining cars began their trips all the way south at the Dorchester–63rd St. stop in the Woodlawn neighborhood. Passengers boarding at the South Wabash station generally waited in a large room on the ground

The North Shore Line's main downtown station at 223 S. Wabash Ave., notable for its decorative signs, was one of the primary gateways to the northern suburbs and Milwaukee. While not a terminus for trains, the station, shown here circa 1940, was home to the company's downtown ticket office and was equipped with a spacious second-floor waiting room. (Shore Line Interurban Historical Society)

floor and went up one flight of stairs to reach the enclosed walkway that protected them from the weather when walking to the platforms.

Travelers could reach Milwaukee from this station's earliest days, but the trip became much faster after the North Shore Line opened its "Skokie Valley Route" in 1926 between Howard Street Station at the north edge of the city and North Chicago. Travel time from Adams–Wabash to Milwaukee dropped to just under two hours. That same year, the North

The North Shore Line's station was connected to the Adams–Wabash "L" station by an enclosed walkway. Stairs within brought patrons to the east platform, which was used by outbound trains to Milwaukee. (Shore Line Interurban Historical Society)

NORTH SHORE LINE ADAMS/WABASH STATION
223 S. Wabash Ave.

A northbound CTA Green Line train departs the Adams/Wabash station on August 19, 2013, only a few months after the 50th anniversary of the demise of North Shore Line. An elevated walkway once linked the North Shore Line's waiting room to this platform, which is still made of wood. (Xhoana Ahmeti photo)

Shore moved the ticket office and waiting room a short distance south into less-cramped quarters at 223 S. Wabash Ave., while maintaining the bridge to the platforms.

This "storefront station" shouldered a heavy burden, accommodating passengers bound for Lake Forest, Mundelein, Waukegan, Milwaulkee and other points on the North Shore Line's 110-mile system. Both comfort and speed improved again after a pair of streamlined *Electroliners*—lightweight trains replete with air conditioning and buffet-lounge service—entered service in 1941. Groups of servicemen stationed at Fort Sheridan and Great Lakes Naval Station added to the wartime surge in traffic.

The North Shore Line's 61 trains through the Adams/Wabash stop in 1950 kept the waiting room and lunch counter busy, but the interurban's financial performance was slipping. A fire the following year destroyed much of the waiting room, which was rebuilt with a streamlined appearance. Then a devastating strike stopped service for 91 days in 1951, during which time the station was closed. Ticket sales fell sharply as automobile ownership increased and highway access improved, particularly after the opening of the Edens Expressway in December 1951.

The railroad abandoned its Shore Line Route from Howard St. to Waukegan in 1955, ending its service to the downtowns of Highland Park, Lake Forest, Winnetka, and other prominent residential communities. By the late 1950s, its deficits had risen to unsustainable levels. The condition of the downtown station worsened, and it was permanently closed when the North Shore Line halted operations on January 21, 1963. The connecting passageway was promptly removed and the storefront space used for other purposes. By the early 1990s, the building had come down to support a major expansion by the Chicago Symphony Orchestra.

Today, the old terminal site is a parking lot that may one day be used for a mixed-use expansion of the Symphony Center concert venue. Adams/Wabash remains one of the most heavily used stations on the CTA system, serving the Brown, Green, Orange, Pink, and Purple Lines. Transit users riding Purple Line Express trains to Howard will experience a trip not unlike that on the North Shore Line years ago, but rides all the way to Milwaukee can be found only in memories. ▪

The wide range of connecting services available at intermediate stops, including the popular Kenosha–Lake Geneva bus service, is vividly shown in this North Shore Line holiday advertisement circa 1924. Trains departed hourly for Milwaukee. The fast Skokie Valley route between Howard Ave. and Waukegan had been built by the time of this ad. (Author's collection)

The South Shore Line's ticket office at Randolph St. Station, shown here circa 1976, typified the terminal's austere and bunker-like qualities—a reputation it deserved because of its location directly below the street. The sign and friendliness of the South Shore's agents, however, helped enliven the passenger experience. (D.W. Davidson photo)

By virtue of its prime location only steps away from North Michigan Ave., Randolph Street Station, known since 2006 as Millennium Station, has been akin to a pawn in a high-stakes game of chess involving real estate development and the creation of cultural amenities. This station's 10 tracks, used by South Shore Line and Metra Electric (formerly Illinois Central Electric suburban) trains, are now entirely enclosed by Millennium Park and modern office-tower development. Although today's station bears little resemblance to that existing at the end of World War II, it has for more than 80 years been the principal terminus for those traveling by train between downtown Chicago and South Bend, IN, as well as many close-in destinations south and southeast of the city

Randolph Street Station was much more significant to long-distance travel 125 years ago. In 1893, however, its owner, the Illinois Central Railroad, moved its terminus for long-distance trains to the much larger Central Station, just over a mile south at Michigan Ave. and Roosevelt Rd. For the next 35 years, Randolph predominantly served suburban trains operating as far south as Richton Park, and the number of passengers handled by the station gradually grew. Construction of a pedestrian tunnel linking the station to the corner of Randolph St. and Michigan Ave. in 1921, and electrification of the route leading into this station in 1926, greatly improved the efficiency and traffic flow.

The South Shore Line began operating its trains into Randolph Street Station in 1926 under a trackage-rights agreement giving it access to the IC route north of Kensington Junction at 115th St., near the southern edge of the city. New platforms were installed on the east end of Randolph Street Station for these trains, which brought throngs of passengers from various points in northern Indiana, including Gary, Hammond, Michigan City, and South Bend.

(Opposite) A South Shore Line train to South Bend, IN, stands at its Randolph St. Station platform immediately south of the 41-story Prudential Building in 1976. In the early 2000s, this part of the station was blocked from sunlight when Millennium Park was built overhead. (D.W. Davidson photo)

RANDOLPH STREET (MILLENNIUM) STATION
151 E. Randolph St.

An Illinois Central commuter train waits at the platform at the north end of Randolph St. Station in 1951. The station's South Water St. entrance, likened to a "chicken coop on stilts," can be seen in the distance. Years later, the construction of the 233 N. Michigan Bldg. and Illinois Center put a roof over this end of the station. (John Fuller photo)

An Illinois Central Gulf commuter train, having just left Randolph St. Station, emerges from subterranean tracks in "the hole" (left), while most of the five upper-level platform tracks (right-center) are occupied by South Shore Line equipment. All of the tracks visible in this 1976 photo are covered by overhead structures today. (D.W. Davidson photo)

RANDOLPH STREET (MILLENNIUM) STATION
151 E. Randolph St.

A lone passenger waits for a South Shore Line train at Millennium Station's Track 13, most of which sits directly underneath Millennium Park, on June 9, 2012. This train will soon make the 88-mile trip to South Bend, IN. (Author's photo)

In 1931, Illinois Central suburban trains were moved to a new five-track facility below ground level—an arrangement that simplified construction of office buildings using "air rights" above the tracks. Arriving passengers could exit at the south end of the platforms to reach Randolph St. or at the north end to reach South Water St. The station had a small and rather austere waiting room above the tracks that was reached by wooden stairways.

When the Prudential Building—one of Chicago's first postwar skyscrapers—was built above the tracks in the early 1950s, the configuration of the station changed again. A giant slab of concrete that supported the 52-story office tower covered the main part of the station. A new waiting room and ticket office were built one floor above the IC tracks and one floor below street level (essentially in the basement of the Prudential). The redesigned station had one more track—a sixth—for Illinois Central trains and the same five for the South Shore.

The public did not receive these changes warmly. Although the station had considerable square footage that included retail shops and a passenger concourse, low ceilings and concrete floors gave the complex a cave-like feel. The terminus proved to be an uninspiring gateway to Chicago for the many passengers who took advantage of Illinois Central's policy of honoring long-distance tickets on suburban trains from Central Station. Recognizing that Randolph St. was in the heart of the city, many made such transfers to save on taxi or transit fare, only to find themselves in an uninspiring windowless station that was difficult to navigate.

The station's image worsened when the roof began to leak, requiring "temporary" plywood walls to be installed. Fixing this problem dragged on for years because it required rebuilding the entire Randolph St. superstructure above. Some likened the station to a bomb shelter, which in fact, was the name of a tavern inside the facility. Making matters worse, the station lacked a clearly identifiable front door or taxi stand.

Randolph Street Station nonetheless remained a transportation workhorse, handling 377 weekday trains in the late 1950s (333 on IC and 44 on the South Shore Line). IC suburban trains at this time operated as far south as Richton Park, IL, a distance of 29 miles, with branches serving Blue Island and South Chicago. Several South Shore Line trains operated 90 miles to South Bend.

Passengers were provided more comfortable and reliable service after Illinois Central introduced sleek two-level "Highliner" cars in 1971. Calls to create a more aesthetically pleasing waiting room, however, went unanswered, triggering allegations that the facility's neglect (as well as the lack of restrooms on the Highliners) was a subtle form of racism and classism. Many African American passengers living in relatively poor South Side neighborhoods bore the brunt of these problems.

Significant changes did not occur until Millennium Park's construction in early 2002. Although the park blocked what little sunlight still reached the platforms, it was a springboard for improvement. An award-winning redesign by Skidmore, Owings & Merrill brightened the station and enlivened its retail and restaurant space. Completely remodeled, it reopened in 2006 with an Art Deco-inspired waiting room and a new name—Millennium Station—as well as a Starbucks Coffee outlet, a Polish sausage purveyor, and other retailers.

Yet, for all the improvements, the station still lacks a clearly identifiable front door or a direct entrance to that famous park after which it is named. This problem has been difficult to overcome due to decisions made during the Prudential Center's construction more than a half century ago. ■

The Shops at Millennium Station, one level above the Metra tracks, see heavy traffic on June 2, 2012. An award-winning design by the architectural and engineering firm of Skidmore, Owings & Merrill transformed the once-dingy, windowless terminal area into "a luminous and cloud-like space." (Author's photo)

WELLS STREET TERMINAL
314 S. Wells St.

The Chicago, Aurora & Elgin Railroad's Wells St. Terminal may have the appearance of being functional in this February 17, 1955 photo, but it had already been closed for more than a year. Crews were beginning to dismantle the platform canopies, visible behind the grand façade. The south end of the Quincy-Wells "L" station—to which the terminal had a direct entrance—can be seen at the lower right. (Krambles-Peterson Archive)

This second-floor view of Wells St. Terminal, taken on June 28, 1927, is one of only a few interior photos known to exist in a publicly accessible archive. Staircases link the waiting room to street level, while the door on the left wall in the distance leads directly to train platforms. (Krambles-Peterson Archive)

Wells Street Terminal, an iconic station on account of its ornate grand façade, was a busy endpoint for trains on several "L" lines and the heavily used Chicago, Aurora & Elgin Railway. The three-level terminal, designed by noted architect Arthur Gerber, had a spacious waiting room and an enclosed walkway linking it to the adjacent Quincy–Wells "L" platforms.

A previous terminal, Fifth Avenue Station, opened at this site in 1904 and gave the Metropolitan "L" (Met) the ability to run certain rush-hour trains into a terminal rather than operating around the congested Loop Elevated. This terminal was equipped with two platforms and four tracks, greatly improving the efficiency of the four Met-operated lines extending west and northwest of downtown.

The Aurora, Elgin & Chicago Railway (CA&E's predecessor) began using Fifth Street Station in 1905. Using the Met's tracks from a junction on the city's West Side, AE&C offered fast and direct service between Chicago and Aurora, a distance of 40 miles, and other points west of the city. Affectionately known as the "Roarin' Elgin" and the "Sunset Line," it relied on an electrified third rail for power on its principal routes. The company became the Chicago, Aurora & Elgin Railroad in 1922.

By the time Fifth Street Station was renamed Wells Street Terminal in 1916, its modest waiting room and platform capacity were ill-equipped to handle the traffic growth. After CA&E came under the control of public-utility magnate Samuel Insull, it finished a massive modernization program that culminated in the opening of a new and much larger Wells Street Terminal in 1927. Like its predecessor, this terminal had four station tracks, all elevated, and two island platforms, but it also had a newly built side platform that could be

Taken from an overhead vantage point and facing east, this view shows several CA&E interurban and "L" trains at a crowded Wells St. Terminal just before rush hour, circa 1915. The train at right, on the south track, is at the station's newspaper loading dock (identified by its slanted roof), while the columns of the terminal building (top) provide a regal backdrop for the "L" structure, most of which was designed in a more utilitarian fashion. (Krambles-Peterson Archive)

WELLS STREET TERMINAL
314 S. Wells St.

A "Roarin' Elgin" train is west of Wells St. Station, on the bridge above Union Station's platform canopies, in 1951. The towers used for dispatching trains and the massive flat-topped Insurance Exchange Building are visible in the distance. (Mark Llanuza collection)

used to load eight-car trains on one of the tracks. The new three-story station, designed by Gerber, had an impressive exterior of terra-cotta and a large ground-floor lobby featuring a restaurant and soda fountain. A "transfer bridge" gave pedestrians direct passage to the Quincy–Wells "L" station.

Trains arrived and departed every few minutes on weekdays through the busy World War II years. During rush-hour periods, trains on the four "L" branches to Garfield Park, Logan Square, Humboldt Park, and Douglas Park originated at the terminal. CA&E also had a robust passenger business linking Chicago to Aurora, Batavia, Geneva and St. Charles, and Elgin and many rapidly expanding western suburbs, including Glen Ellyn, Lombard, and Wheaton.[28] The *Cannonball*, the CA&E's only eight-car train, arrived and departed from the lengthy side platform.

Wells Street Terminal's transportation role diminished sharply after the war. The city's "L" routes were transferred to the CTA in 1947, which had little interest in stub-end stations such as this one (i.e., those with tracks that dead-ended). The privately owned CA&E, meanwhile, could not cope with mounting losses in the face of rising automobile competition. As late as 1950, 61 CA&E trains used the aging station each weekday (with only slightly fewer on weekends), but the future was increasingly cloudy as construction of the Congress Expressway, which required the demolition of the Garfield "L," moved forward.

Over the next several years, Wells St. moved ever closer to its ultimate fate. Direct "L" service to Humboldt Park ended. Douglas and Garfield "L" trains were rerouted in 1951 to the brand-new Milwaukee–Dearborn subway. CA&E was eager to vacate the facility as well. Rather than use a replacement route in the middle of the Congress Expressway and a proposed temporary route to the terminal, the "Roarin' Elgin" ceased operations into downtown Chicago. Wells Street Terminal saw its last train depart on September 20, 1953. Passengers headed to the western suburbs formerly served directly by the interurban were now expected to ride CTA trains to suburban Forest Park and make an across-the-platform transfer to CA&E trains to complete their journey. As would be expected, traffic sharply fell.

Wells Street Terminal sat unused for several years before the top two floors were demolished as part of the construction of Wacker Dr. in 1955. Two of the stub tracks in the terminal were connected to the Loop Elevated system, which was used until the Congress rapid-transit line opened.

Abutments from the bridge over the Chicago River leading to the station can still be seen between Jackson and Van Buren streets along the west bank of the river. A parking deck built on the old station site reveals the bygone terminal's location by virtue of being set back several feet from the Wells St. sidewalk. An electrical substation, once behind the terminal, survives to provide power to the CTA. Nothing else remains. ■

This Chicago, Aurora & Elgin Railroad system map shows the range of routes served by the "Loop Terminal" (Wells St. Terminal) as it appeared in its February 2, 1936, timetable. The slogan—"The Direct Way to and from the Loop"—was used to differentiate CA&E's services from competing railroads' commuter trains, most of which deposited passengers only on the fringe of the Loop district.

STEAMSHIP LANDINGS
Terminal Town

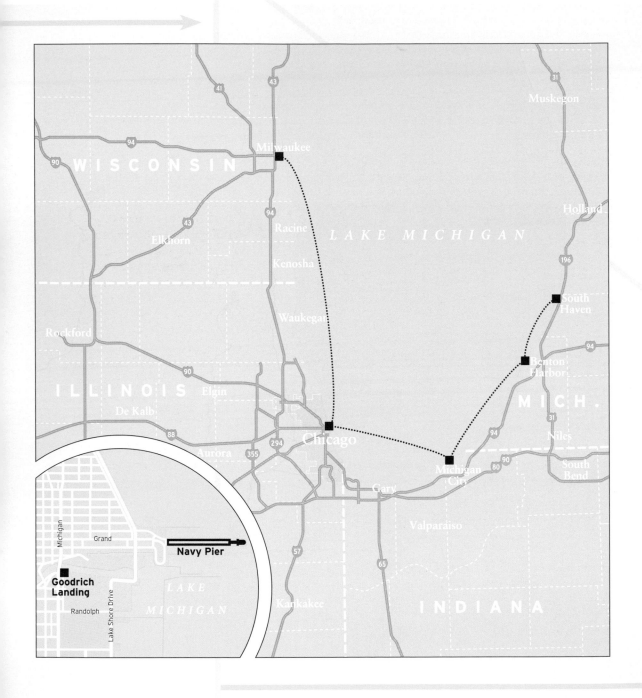

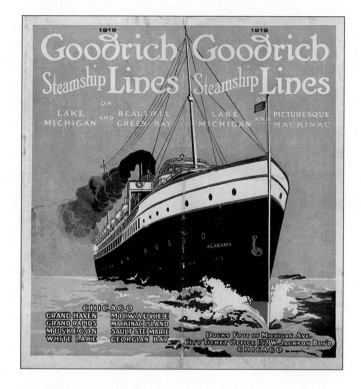

A Goodrich Steamship Lines advertisement promotes its sailings to many Great Lakes cities from "docks at the foot of Michigan Ave." The company had yet to relocate its main docking area to the newly completed Municipal Pier. (Wisconsin Historical Society, WHi-5763)

A proliferation of steamship landings along the Chicago River was the source of blight, congestion, and pollution throughout much of the early twentieth century. Wharves lined the riverbank in a haphazard fashion, all the way from present-day Columbus Dr. to Wolf Point, where the North and South branches diverge from the Chicago River's main stem. Passenger and freight activities were often intermixed, creating a sense of commotion and disorder.

The orderliness of river traffic significantly improved with the opening of Municipal Pier in 1916. By then, however, steamship travel had begun its slow, inexorable decline. The demand for waterborne passage continued to drop as growing automobile ownership, ever-faster rail service, and other factors accelerated the pace of passenger travel.

By the late 1930s, steamship service on the Great Lakes was a mere shadow of the previous decade's levels. Ravaged by the Great Depression and technological change, the little service that remained was provided at only two locations, Goodrich Landing and Navy Pier, and was mostly oriented toward vacationers bound for Milwaukee and the eastern shore of Lake Michigan. Despite this, steamships remained part of the region's transportation scene through the wartime years and until the end of the summer season in 1950. They are fondly remembered by many area residents.[29] The sites of both terminals, now redeveloped, serve as stops for water taxis and pleasure cruises. ■

(Opposite) This map shows the routes served by scheduled passenger steamship lines at various times between 1939 and the end of service in 1950. By the start of World War II, only this skeletal system remained, linking Chicago to Milwaukee, Benton Harbor, Grand Haven, and Michigan City. As the inset map shows, these services operated from one of two downtown terminals: the Goodrich Landing at the foot of Michigan Ave. along the Chicago River, and Navy (Municipal) Pier. The street grid depicted dates to 1950–51.

GOODRICH LANDING
Michigan Ave. & Chicago River

Goodrich Landing, situated on the south side of the Chicago River immediately east of Michigan Ave. (a site directly across the street from today's Hyatt Regency Chicago at 151 E. Wacker Dr.), served generations of travelers bound for Benton Harbor and St. Joseph, MI, Milwaukee, and other Lake Michigan ports. This was the last steamship landing in Chicago with scheduled service to a destination beyond the metropolitan region.

Through the early twentieth century, the Goodrich Co. was one of Chicago's largest and best-known steamship providers. Passengers liked the company's modern vessels and its wharf's proximity to streetcar and rapid-transit routes. Advertised as being "at the foot of Michigan Ave.," it was only three blocks from the Randolph & Wabash "L" stop and even closer to Randolph Street Station. Goodrich Landing also handled a great deal of freight transferred between steamships and trains using Illinois Central's nearby South Water St. rail yards, which also served the New York Central, Nickel Plate Road, and Chesapeake & Ohio railroads.

The number of sailings from Goodrich Landing diminished after the opening of Municipal Pier (later Navy Pier) in 1916, when the *Theodore Roosevelt* and other Goodrich Co. ships were transferred to this massive new facility. Still, the wharf remained a busy place, despite the company's bankruptcy in 1932. The *Theodore Roosevelt* returned to the wharf around 1935, in part because passengers (many of whom arrived

The SS *Theodore Roosevelt* prepares to sail from the former Goodrich Landing on the south bank of the Chicago River in the shadow of the Loop's skyscrapers in 1936. This vessel, calling at Navy Pier under charter to Goodrich between 1927 and 1932, was profitable on Michigan-bound excursions but less so on the run to Milwaukee. The smaller vessel near the stern of this steamer also appears to belong to Goodrich. The wharf site, just east of Michigan Ave., was directly in front of today's Hyatt Regency Chicago at 151 E. Wacker Dr. (John F. Humiston photo)

downtown by bus, streetcar, or the "L") preferred the location to the more outlying Navy Pier. Large numbers of seasonal passengers embarked on this steamship for St. Joseph and Benton Harbor, with service extended to South Haven, MI, in 1936.

Over the next decade, however, Goodrich Landing had only sporadic service available to travelers apart from summertime cruises on the lake. The *Theodore Roosevelt* stopped sailing from Chicago in 1940 but returned to the wharf for one final season in 1946, when it resumed its old Benton Harbor run under the tutelage of the Cleveland & Buffalo Steamship Co. A smaller vessel, the *City of Grand Rapids*, also returned to the wharf in 1942, operating to St. Joseph and Michigan City, IN. After the 1943 season, the *City of Grand Rapids* began operating over the more picturesque Chicago–Milwaukee route. This lasted until 1947, when this ship was redeployed on the Benton Harbor route. This service ended at the end of the 1950 season. Without fanfare, the era of regularly scheduled steamer service from Chicago drew to a close.

All traces of Goodrich Landing have vanished. Much of it was likely swept away during the 1963 extension of Wacker Dr. east of Michigan Ave. and the construction of the Illinois Center, a vast mixed-use development on former Illinois Central Railroad property. Today, however, several cruise companies dock at the southeast corner of the Michigan Ave. Bridge, virtually the same spot as the old Goodrich Landing. Other boats use the Wendella Company's wharf on the opposite side of the river. ■

NAVY PIER
600 E. Grand Ave.

Navy Pier, while no longer a terminal for marine passenger vessels on scheduled trips to points outside the region, remains a popular embarkation point for cruise operators. This photo shows the *Odyssey*, operated by Entertainment Cruises, near the spot once used by massive Goodrich steamers. A water taxi is moored in the distance. (Entertainment Cruises)

In the early part of the twentieth century, when the Great Lakes steamship era was in full bloom, Chicago heralded Municipal Pier as the solution to congestion on the Chicago River that was "universally considered intolerable."[30] The Burnham Plan envisioned the pier, and another to be built farther south, providing modern docking space for passenger vessels and also serving as a public gathering place.

The pier's completion in 1916 brought great optimism to transportation planners. Officials throughout greater Chicago saw the massive new facility, designed by architect Charles Sumner Frost, as an innovative feat that would finally permit consolidation of scheduled passenger and excursion operators at a single marine terminal. Soon after its opening, both the Northern Michigan and Roosevelt Steamship companies moved their operations to the pier. In 1926, the powerful Goodrich Co. followed suit, occupying a spot on the pier's southwest corner. A new streetcar service operating to the pier improved access for people arriving from the city side.

Despite all its promises, Municipal Pier fell far short of expectations. Its distance from the city's produce markets—a key source of revenue for steamship operators—was deeply problematic, and this shortcoming only grew more severe as the fruit and vegetable trade gradually migrated south of downtown. Worse still, many passengers considered the location to be inconvenient. Most preferred the berthing areas closer to the central part of the city that were only a few blocks from the Loop Elevated system.

Underused and costly to operate, the 3,300-foot pier became something of a white elephant amid concerns that it would never see as much commercial shipping as its promoters envisioned. Making matters worse, the demand for steamship service fell in response to automobile and railroad competition. "The pier simply came too late," concluded George W. Hilton in *Lake Michigan Passenger Steamers*.[31]

Gradually, customers came to regard Great Lakes passenger steamships as a form of leisure rather than practical transportation. This was especially true for summer sailings to nearby Benton Harbor, MI, Michigan City, IN, and other cities near the south end of Lake Michigan that could be reached by rail in considerably less time. During the winter months, ice conditions on the lake caused a cessation of this shipping activity.

The SS *City of Grand Rapids*, a mainstay on the Chicago–South Haven, MI, route, is shown at the Goodrich berthing area near the southwest corner of Navy Pier circa 1930. This 290-foot-long steamer was built in 1912 for the Graham & Morton Transportation Co., which merged with Goodrich in 1925. Subsequent owners, also providing Lake Michigan service, included the Chicago–Milwaukee Steamship Line (1937) and the Cleveland & Buffalo Steamship Co. (1942). The vessel was retired in 1951. (Chicago Transit Authority)

NAVY PIER
600 E. Grand Ave.

Navy Pier (renamed from Municipal Pier in 1927), measuring 3,300 feet in length, was widely regarded as a modern-day maritime marvel when it opened in 1916. The massive terminal was underutilized from the start, however, and by the time of this photograph, circa 1933, scheduled steamship service had reduced to a trickle. Two large boats are docked at the Goodrich Transit Co. berthing area (right), but the "Goodrich Boats" sign on the front of the terminal appears to be in the process of being painted over, likely as a result of the company's December 1932 bankruptcy. (Shore Line Interurban Historical Society)

The Great Depression brought financial ruin to much of the Great Lakes passenger steamship business. Although many ships stopped operating, two Navy Pier stalwarts, the *Theodore Roosevelt* and *City of Grand Rapids*, continued to sail. The *Theodore Roosevelt* operated to Michigan City, while the *City of Grand Rapids* operated primarily to Milwaukee and St. Joseph, MI.

Navy Pier's role diminished further when the *Theodore Roosevelt's* berthing area was moved to the Goodrich Landing, a wharf adjacent to the Michigan Ave. Bridge (near today's Illinois Center), around 1935. This left only the *City of Grand Rapids*, which sailed from Navy Pier to Milwaukee through 1941, when it was transferred to the routes linking Navy Pier with St. Joseph and Michigan City. By this time, the pier's days as an intercity passenger terminal were numbered. The *City of Grand Rapids* was moved to Goodrich Wharf in 1942, largely due to expanding naval operations at the Pier during World War II, ending scheduled passenger operations at the lakeside facility after 26 years.[32]

A lengthy debate about the future of Navy Pier ensued. After serving as the

A 1930 Goodrich Steamship Lines advertisement offers a dramatized perspective of one of its steamers leaving Navy Pier and the city skyline, enjoyed by its passengers. The vessel may be the *City of Grand Rapids*, a Lake Michigan stalwart through the end of the 1950 season. (Wisconsin Historical Society, WHi-5787)

University of Illinois' Chicago campus from 1946 to 1965, it became a trade show, exhibition, and festival site through the 1980s. Following a multi-year reconstruction, it reopened in 1995 as a festival marketplace that earned accolades as one of the Midwest's leading tourist attractions. A new generation of pleasure-cruise ships sailing the Great Lakes began docking at Navy Pier in increasing numbers. The tourist traffic generated by the redesigned pier spawned water-taxi service to points along the Chicago River and Museum Campus, thus beginning a new era of scheduled waterborne passenger service.

Proposals for establishing a Chicago-based high-speed hovercraft route across Lake Michigan have occasionally emerged, but no scheduled service beyond the water taxis has materialized. ∎

II. OUTLYING STATIONS
Terminal Town

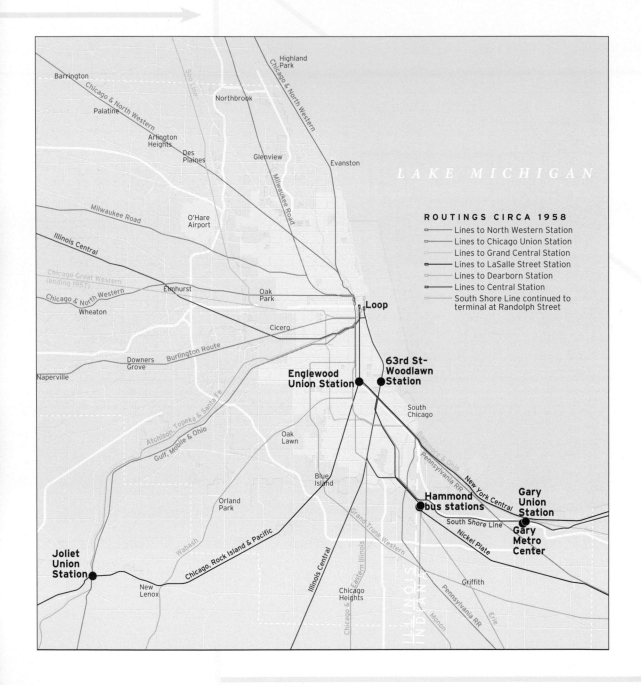

ROUTINGS CIRCA 1958
- Lines to North Western Station
- Lines to Chicago Union Station
- Lines to Grand Central Station
- Lines to LaSalle Street Station
- Lines to Dearborn Station
- Lines to Central Station
- South Shore Line continued to terminal at Randolph Street

An Amtrak Turboliner train bound for St. Louis on the former Gulf, Mobile & Ohio Railroad calls at Joliet Union Station on May 3, 1976. Passengers arriving from points west of Chicago could transfer here to trains running to central Illinois, shaving more than 70 miles off their trip. (Verne Brummel photo)

Passengers traveling through the Chicago region years ago could often save time, effort, and money by transferring at a station outside of the central city. Such connections also could eliminate the hassle and expense of transferring between downtown bus and railroad stations.

Opportunities for timesaving transfers were particularly abundant on the South Side. This was a legacy of downtown Chicago's location too far north of the southern tip of Lake Michigan to be optimal for many passenger connections. Although being located at the mouth of the Chicago River was advantageous for waterborne vessels, the city's downtown was farther up the lake's shoreline than desirable for many long-distance bus and train passengers wanting to move across the Midwest—or the country—in as straight a line as possible.

Lake Michigan's presence made it inevitable that a variety of connecting hubs would emerge in the southern part of the region. Railroad companies building from eastern points, for example, could not follow a direct path to the city center due to the impossibility of bridging

(Opposite) This map shows the most significant points where passengers passing through the Chicago region could make timesaving connections and thereby eliminate the need to transfer at downtown terminals. Using these locations could save passengers anywhere between 13 and 80 miles of travel. Illinois Central's 63rd St.–Woodlawn Station, Englewood Union Station, and the various bus stations in Hammond, IN, were especially heavily used for intercity bus and train transfers. Today's Amtrak stations in Glenview, Homewood, La Grange, and Naperville are important connecting points between long-distance and commuter trains, but are less significant as sites of intercity transfers.

this freshwater body. The eastern roads instead threaded their way through the dense network of track arcing through northwestern Indiana and approaching downtown through working-class South Side neighborhoods.

Few railroads could afford to build their own dedicated routes into Chicago's downtown stations. Most used the tracks of other mainline railroads, such as the Illinois Central, or "terminal railroads," such as the Baltimore & Ohio Chicago Terminal (which operated into Grand Central Station) and the Chicago & Western Indiana Railroad (owner of Dearborn Station), which provided efficient links to downtown stations for several different railroads. The terminal railroads were akin to high-voltage conduits running between the region's outskirts and terminals in the central city.

Despite the plethora of junctions that emerged southeast, south, and southwest of downtown, relatively few became significant transfer points for long-distance travelers. The five railroad stations along 63rd St., however, were among the most notable.[33] The Illinois Central station at 63rd and Dorchester, and Englewood Union Station on 63rd between State and Wentworth, were true connecting hubs that allowed passengers to shave more than ten miles from their trips. Some passengers even made connections *between* these South Side stations by using the 63rd St. streetcar line to eliminate the need for transfers between terminals in downtown Chicago.

Passengers could also save time or money by transferring at stations even farther outside of the city, although the opportunities to do this were comparatively few. The union stations in Hammond and Gary, IN, and Joliet, IL, were impressive places, but they were only marginally important connecting points for travelers on intercity train trips through Chicago. All lacked the types of transfer connections needed to be important hubs akin to those on 63rd St.

Motor-coach travelers had more opportunities to make timesaving connections at outlying stations than their train-riding counterparts, with the bus depots of Gary and Hammond being particularly important crossroads. As hard as it may be to imagine today, buses traveling between the Windy City and Memphis, New Orleans, and St. Louis, routinely stopped in Hammond, IN to pick up connecting passengers arriving from points east of the city.

Making transfers in these outlying communities could nonetheless be confusing and cumbersome, especially when it required transferring between stations. Many passengers walked, rode buses or streetcars, or hailed cabs to transfer between stations, particularly in Gary and Hammond, where bus and train service was spread out over a half dozen or more stations, some of which were over a mile apart. None of the aforementioned satellite cities brought bus and trains under the same roof until the opening of the Gary Metro Center in 1986, and even that station was served by only one rail line.

Today, almost all long-distance passengers making connections between buses or trains within the metropolitan region do so in downtown Chicago. Outside of downtown, only the Gary Metro Center caters to passengers making transfers on trips that neither originate or terminate in the region. Sadly, most terminals discussed on the following pages have either disappeared or now exclusively serve travelers making public-transit trips within the metropolitan region. ■

(Opposite) Buses operated by Greyhound and Indian Trails stop at Gary, IN, in August 1974. Passengers on the latter company's buses from western Michigan destined for the East Coast often transferred in northweste rn Indiana—either Gary or Hammond—to save themselves the time and expense of traveling through downtown Chicago. (Mel Bernero photo)

The Pennsylvania Railroad's premier train between Chicago and New York City—the *Broadway Limited*—pauses at Englewood Union Station on April 21, 1965. Passengers on this all-Pullman run, and other trains from the East Coast, could make timesaving transfers here to Rock Island trains heading west. The station (right center) appears to be in good condition, despite the challenges facing the surrounding neighborhood. (Marty Bernard photo)

An E6A locomotive is at the head of a southbound Rock Island commuter train at Englewood on April 21, 1965. The conductor can be seen boarding in the distance. (Marty Bernard photo)

ENGLEWOOD UNION STATION
63rd St. & State St.

Englewood Union Station's arrival and departure board, shown here circa 1964, still showed extensive service on the New York Central, Pennsylvania, and Rock Island Lines, as well as an eastbound and westbound Nickel Plate Road train. The amount of service, however, quickly dwindled before the last of service at Englewood Union Station came to an end in the 1970s. (John Fuller photo)

For passengers on long-distance trains, Englewood Union Station was unquestionably the region's most important transfer point outside of downtown Chicago. Situated at the crossing of three busy passenger routes—the Rock Island, New York Central, and Pennsylvania Railroad (PRR) main lines—Englewood eliminated the need for many passengers to make time-consuming transfers at downtown stations, trimming as much as 90 minutes from their journeys. A fourth railroad, the New York, Chicago & St. Louis (widely known as the Nickel Plate Road), also served the station using New York Central's tracks. In addition to its connecting traffic, Englewood was an important point of embarkation for South Side residents and business travelers.

Opened in 1898 and located on 63rd St. between State and Wentworth, this brick-and-stone structure had many of the qualities of a downtown station, boasting a ground-floor telegraph office, a taxi stand, a spacious waiting room, a newsstand, and ticket offices one floor up at track level. The baggage room was located in a separate building. Streetcars and buses stopping in front of a lower-level passageway made the station relatively easy to reach.

Englewood Union Station's four railroads operated from three sets of platforms that formed a triangle on the property. Englewood was the first stop for many trains leaving downtown Chicago and the second-to-last stop for many destined for the Windy City. The roster of trains rivaled that of Chicago's great downtown stations. The PRR's eastbound *Broadway Limited* and New York Central *20th Century Limited*, both of which operated between Chicago and New York, at one time stopped at Englewood almost side by side and departed simultaneously, much to the delight of photographers seeking to capture the majesty of these great trains.

Nickel Plate and Rock Island trains reached Englewood over New York Central tracks from LaSalle Street Station. After arriving at Englewood, Rock Island trains headed southwest toward Joliet, while the New York Central and Nickel Plate veered southeast to reach northwestern Indiana. PRR had its own tracks from Chicago Union Station through Englewood—a route that largely paralleled New York Central, east of this famous junction.

The range of services offered at Englewood was enormous. In 1942, Englewood boasted 101 long-distance trains and 24 commuter trains, with PRR accounting for 31 of the former. Englewood was the only Chicago station where passengers could board trains to every major city along the Eastern Seaboard. Passengers could depart for the New York City region on three different railroads at Englewood. Remarkably, in 1946, travelers from Englewood could directly reach 52 of America's 100 largest cities, more than from any other station in Chicago and all but one terminal in the United States, St. Louis Union Station, which had service to 60 cities.[35]

Thousands of passengers made connections at Englewood annually on trips between points east and west of Chicago. A particularly large number of westbound passengers made connections to the Rock Island's famed *Rockets* operating to Des Moines, Denver, and other points, as well as its *Golden State*, which operated to Phoenix and Los Angeles. Passengers in earlier times also enjoyed direct service to the Twin Cities.

A public notice posted at Englewood Union Station informed users of the bad news—the station building would close on April 9, 1969. After that date, passengers no longer had access to a station agent and had to wait outside. Englewood remained a station stop for several more years. (John Fuller photo).

An eastbound New York Central train from LaSalle Street Station rounds the curve into Englewood Union Station on April 21, 1965. The Rock Island tracks can be seen branching in the distance. (Marty Bernard photo)

ENGLEWOOD UNION STATION
63rd St. & State St.

Englewood Union Station routes, shown here on January 1, 1949, offered an extraordinary number of timesaving connecting opportunities for passengers moving between points on the eastern-oriented New York Central, Nickel Plate Road, and Pennsylvania Railroad, and the western-oriented Rock Island Lines. Travelers could spare themselves the hassle of transferring between downtown terminals, and could often shorten the duration of their trips by making transfers at this South Side junction.

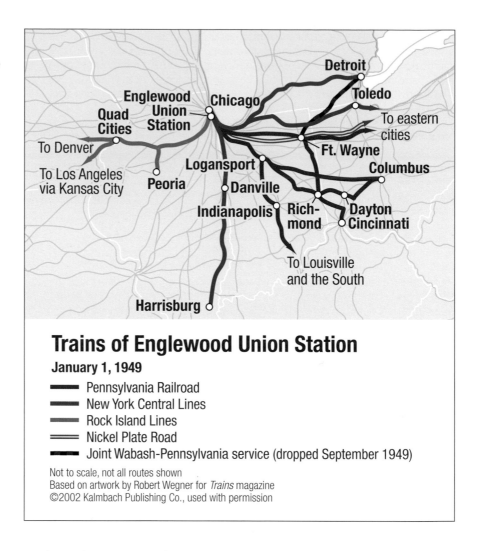

Trains of Englewood Union Station
January 1, 1949

━━━ Pennsylvania Railroad
━━━ New York Central Lines
━━━ Rock Island Lines
═══ Nickel Plate Road
━━━ Joint Wabash-Pennsylvania service (dropped September 1949)

Not to scale, not all routes shown
Based on artwork by Robert Wegner for *Trains* magazine
©2002 Kalmbach Publishing Co., used with permission

Englewood Union Station lost much of its vitality as the surrounding neighborhood depopulated and suffered large-scale divestment in the late 1960s. White flight and rampant crime created such problems that travelers increasingly avoided using the station. Another difficulty after the demise of streetcar service was the lack of a direct connection to the city's "L" system, creating a sense of isolation.

The termination of the last of the former Nickel Plate passenger trains in 1965, and the discontinuing of Rock Island's *Golden State* in early 1968, further reduced the number of passengers using the station. By the end of 1968, the number of intercity trains calling at Englewood had fallen to 27. Some tracks were removed after New York Central and Pennsylvania merged to form Penn Central in February of that year, after which almost all of that new carrier's trains into Chicago were rerouted into Union Station over the former PRR.

Englewood's waiting room closed and the ticket office ceased being staffed in 1969. Penn Central trains made their last stops at the inception of Amtrak on

May 1, 1971, and the new national carrier opted not the use the station. Rock Island's remaining long-distance trains ceased using the station several years later, and its commuter trains began skipping Englewood in the late 1970s.

The station, windows broken and covered with graffiti, was demolished soon after, and only scattered remnants exist today—most prominently, the massive concrete viaduct near 63rd and State streets. In 2011, federal funds were made available to build a new track overpass to eliminate the busy Englewood crossing, which was by this time used by Amtrak and Metra passenger trains, as well as Norfolk Southern and CSX freights. This project is slated for completion by 2016. ∎

A Pennsylvania Railroad commuter train linking Valparaiso, IN, to Chicago Union Station—the second of two weekday inbound trips—stops at Englewood Union Station on April 21, 1965. In this eastward-facing view, the Rock Island main line is in the foreground. Englewood was the most significant timesaving connecting point in the region for train travelers. (Marty Bernard photo)

GARY METRO CENTER
200 W. 4th Ave., Gary, IN

An eastbound South Shore Line train stops at Gary Metro Center on Labor Day 2012. The pedestrian bridge visible to the left of the train links the bus-loading area (left center in the distance) with the train station—a rare example of a Chicago-area terminal designed to provide connectivity between different modes of transportation operating from disconnected locations. (Author's photo)

Gary Metro Center serves not only as an important station for residents of Indiana's Calumet region; it offers numerous opportunities for timesaving connections for those passing *through* the Chicago metropolitan area. For more than a decade, this station had direct bus service to more major cities in the United States than any other "outlying" station in the metropolitan region. Today, Metro Center remains an important station and transfer point for the local bus operator, two intercity bus providers, and the Northern Indiana Commuter Transportation District (South Shore Line).

Support for building this station gathered momentum as the economic downturn that crippled Gary in the early 1980s made consolidation of the city's intercity bus, commuter rail, and local transit services an important priority. Merging most of the city's numerous stations into one was seen as a way to enhance the passenger experience and make security easier to provide. Having trains and buses converge at one point facilitated passenger connections, which was particularly advantageous for those on long-distance trips. Travelers going from Benton Harbor, MI, to South Bend, IN, for example, could make bus-train connections in a few minutes instead of making the cumbersome transfer from one station to another.

The result was Adam Benjamin Metro Center, an intermodal hub that replaced much of the aging South Shore Line station at 280 Broadway and all of Union Bus Station at 5th & Adams streets, both of which had more than 40 daily arrivals and departures in 1985. Built by the city of Gary, the center was seen as instrumental to an ambitious downtown improvement plan. The South Shore tracks were elevated above street level to eliminate a dangerous crossing on Broadway St., and a 200-foot pedestrian bridge linked the main station building to the new platform.

Much fanfare surrounded the center's opening in 1986. Its close proximity to the neighboring Genesis Convention Center was touted as contributing to a revival in downtown Gary. Long-distance passengers had access to waiting rooms on two levels and ticket counters for Greyhound and the South Shore Line. Parking for commuters was ample, and buses arrived and departed every few minutes from a bus bay on the south side of the station.

Most Greyhound buses traveling along Lake Michigan's southern edge stopped here, giving Gary express service to Atlanta, Detroit, Miami, New York City, and Washington, D.C. A steady flow of travelers made connections here on trips between Michigan's Lower Peninsula (heavily served by both Greyhound and Indian Trails Bus Lines) and points in the East and South. Such connections saved passengers the time and hassle of traveling through Chicago, as it shaved 62 miles off their trip.

Gary Metro Center, shown in 2012, has ground-floor bus facilities (visible at left) directly below an upper-level waiting room for rail passengers. The station, which is adjacent to an attractively landscaped park, continues to generate extensive intercity bus and rail traffic, although its sprawling design makes upkeep difficult. (Author's photo)

Amtrak trains, however, were notably absent, as that carrier's inclusion was deemed prohibitively complicated and expensive. The carrier's Chicago–Detroit trains, consequently, whisked by on a high embankment several hundred feet north of the South Shore tracks, and stopped instead at Amtrak's Hammond–Whiting station, approximately 12 miles to the west of Gary. Amtrak's *Broadway Limited* and its trains from Chicago to Valparaiso, IN, stopped at a station several miles southwest of Metro Center at 5th and Chase streets.

Significant improvements made to Metro Center in 1989 expanded its office facilities for Greyhound and Hammond Yellow Coach lines, a competitor to the South Shore Line that operated three commuter routes from the station. By the mid-1990s, however, Yellow Coach had discontinued service and the image of Metro Center—like that of Gary itself—began to sink. The station's rugged, reinforced concrete walls gave it a dreary, bunker-like appearance. The

The station sign on 5th St. in front of the Gary Metro Center suffers from neglect, as evidenced by the missing panels. All three of the center's primary carriers, the Gary Public Transit Corp. (top), South Shore Line, and Greyhound, however, still have their logos on display. (Author's photo)

Passengers board an eastbound
South Shore Line train bound for
South Bend from the Gary Metro
Center's island platform in September
2012. (Author's photo)

surrounding streetscape deteriorated as the city's finances were stretched to the limit. Retailers fled the area adjacent to the terminal, and Metro Center, like much of downtown Gary, became dogged by rampant crime.

Despite this, the passengers kept coming and, as late as 2000, the station still had 28 intercity bus operations daily. The next decade, however, was difficult. Greyhound eliminated a route to Fort Wayne, IN, in 2005 and then eliminated its direct service to Boston and Florida. Several local transit routes were cut due to funding shortfalls in 2010. Nevertheless, the station continued to serve as a hub for connecting passengers. Presently, Metro Center has 22 intercity bus departures—the most of any station in the region outside the Chicago city limits—and several dozen daily South Shore Line trains. Metro Center today is the last remaining transfer point significant to both intercity bus and rail travelers in the metropolitan region that is outside of Chicago's city limits. ■

GARY UNION STATION
185 Broadway St., Gary, IN

Gary Union Station, hemmed in by the embankments of the Baltimore & Ohio main line (at right) and New York Central's "Water Level Route" (at left), appears virtually unchanged since its opening, circa 1965. Cast-concrete walls, a relatively new innovation when the station was built, gave the impression of being made of stone. The roadway between the terminal and the B&O embankment links Broadway St. (bottom) to a mail and express building at the rear of the complex. (Julian Barnard photo, B&O Historical Society Collection)

For generations, Gary Union Station was the most prominent railroad station in northwestern Indiana. By virtue of serving two important carriers, the Baltimore & Ohio and New York Central railroads, it had direct service to every major city in the Northeast.[36] Passengers disembarked here, not only to reach points in Gary and its immediate vicinity, but also to transfer to buses and other trains.

Promoters of Gary's steel industry in the early 1900s considered a railroad station that was commensurate with the city's growing wealth and commercial importance to be an utmost priority. E.H. Gary, chairman of United States Steel Corp., oversaw investments that allowed for the relocation of several railroad lines near the lakefront, and the creation of a new station between the newly elevated tracks of the Baltimore & Ohio and Lake Shore & Michigan Southern (the latter company being a New York Central affiliate).

The station's location on Broadway Ave.—one of Gary's busiest streets—gave it an almost iconic status from the moment it opened in 1910. Architect M.A. Lang's two-story Beaux-Arts design featured a large skylight and elegant staircases leading to the tracks. Baltimore & Ohio trains stopped at platforms on the south side of the station, while those of the Lake Shore & Michigan Southern used the north side.

By 1928, 30 long-distance trains stopped at Gary Union Station daily. New York Central offered the most extensive service and accounted for about two-thirds of the departures, including the famed *20th Century Limited*. Baltimore & Ohio's trains, among them the widely used *Capitol Limited* and *New York–Chicago Special*, also provided express service to the East Coast by way of Washington, D.C.

Upon arrival in Gary, passengers fanned out on foot to reach connecting transportation services, including the electrified interurban South Shore Line, which had its main station in Gary only a few hundred feet away. Others made the one-block walk to the front gate of the U.S. Steel complex, where Gary Railways had its main terminal. As late as 1930, this carrier operated a 71-mile network of trolley and interurban routes through the region.

Despite Gary Union Station's extensive service—it had 32 daily trains in 1942—it was not a highly effective connecting hub for long-distance travelers. Most trains operated on an east-west axis. All westbound trains terminated in downtown Chicago and no eastbound train reached farther north than upstate New York, or farther south than Washington, D.C. As a result, there were few opportunities to make transfers in Gary that didn't involve reversing the passenger's direction of travel, thus limiting the number of connecting opportunities within the station. More connection opportunities would have been available had the Pere Marquette Railway's trains, which used the New York Central through town and carried particularly large numbers of people to and from Grand Rapids, MI, stopped, but its trackage-rights deal precluded this. The most attractive connections for long-distance travelers, consequently, generally required transfers to one of the four other railroad stations in the community or to one of the two major bus stations.

Gary Railways abandoned the last of its rail operations in favor of buses in 1947 and the number of passengers using Union Station gradually fell. More favorably, however, trains that had once whisked through Gary without stopping began to call on the station, partially due to the city's rising population. In 1956, the station had 36 daily departures—four more than it had during the war. The following year, trains operating between Chicago and Detroit that had previously used a more southerly (former Michigan Central) route through Gary began using the New York Central mainline through the station.[37]

But the station's best days were long in the past. The Indiana Toll Road's construction immediately south of the depot in 1957 not only made auto travel faster and easier; it isolated the station from the rest of downtown Gary. The Toll Road's hulking pylons elevating its traffic lanes above Broadway St. created a psychological barrier between the station and retail areas nearby.

Massive reductions in the steel industry in the 1960s and 1970s sent Gary's economy spiraling downward, and race relations gradually took a turn for the worse. Although the station still had a dozen trains at the start of 1971, its days were numbered. When Amtrak began service in May of that year, the station—which was costly to heat and had little parking, as it was hemmed in by railroad tracks—was deemed unsuitable and replaced with a new and more modest Amtrak station 12 miles west at Hammond–Whiting.

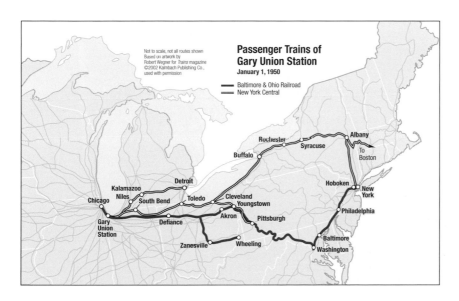

The Baltimore & Ohio and New York Central were almost perfect complements for passengers traveling from Gary Union Station to the East Coast, with the former specializing in points south of New York and the latter serving more northerly destinations. The linear configuration of routes, however, limited the station's effectiveness as a self-contained hub. Nevertheless, many traveling between Niles and other Michigan points and eastern cities made connections at the station, while others disembarked to make transfers to Gary's other stations.

GARY UNION STATION
185 Broadway St., Gary, IN

Both the Baltimore & Ohio platforms—which were equipped with lights but lacked a canopy—and the deteriorating terminal building were woefully underutilized by the time of this 1969 photo. Passengers anticipating the arrival of their train usually walked under the bridge linked to the station building (visible at bottom left) and ascended the staircases to the north or south platforms. (John Fuller photo)

The fate of the enormous old station now hung in the balance. After briefly being used for an automobile salvage operation, it was simply abandoned. Vandals broke many of the windows, portions of the roof collapsed, and harsh winters took their toll on the terra-cotta façade. The station platforms and canopies were removed by the railroad companies, while other parts of the station gradually withered, largely hidden from view by the railroad embankments that flanked it.

Even the possibility of a passenger-rail revival in northwestern Indiana brought little hope that this facility would see trains again, despite the fact that many Amtrak trains passed it every day on the very rail lines for which this station was built to serve. Gary Union Station is on neither the most likely future high-speed route to Indianapolis nor the envisioned potential 220 mph route across northern Indiana linking Chicago to Cleveland and Detroit.

This chalkboard in Gary Union Station still listed nine New York Central daily arrivals and departures in 1969, a year after the NYC and Pennsylvania Railroads had merged to become Penn Central. Nos. 27/28 were remnants of the *New England States*, a Chicago–New York/Boston train that, like most other former NYC trains, by this time operated without a name. No. 356, a Chicago–Detroit train, had formerly been the *Twilight Limited*. (John Fuller photo)

The station, now something of a white elephant, has come to symbolize Gary's economic woes. Too big and awkwardly designed to be renovated, and on a site too narrow to be equipped with sufficient onsite parking to be retrofitted for another use, some believe its fate is all but sealed. ■

Gary Union Station's dismal condition is laid bare in this contemporary photograph. Broken windows and a partially collapsed roof have exposed the interior to the elements for many years. (Author's photo)

HAMMOND'S GREYHOUND
51-52 S. State St., Hammond, IN

A passenger waits outside a Gold Star Line bus at the Schappi bus station in Hammond, IN, in 1963. This station at Hohman Ave. and Willow Ct. was one of a cluster of stations used by the Indiana city's bus lines. Gold Star's Hammond–Joliet, IL, route allowed many passengers to avoid more circuitous trips through downtown Chicago. (Johnnie Williams photo, Krambles-Peterson Archive)

Little known to the region today, the intercity bus hub in Hammond, IN, was for generations second in importance only to Chicago in the metropolitan region. Situated near the southern tip of Lake Michigan—an almost ideal spot for long-distance transportation through the Midwest—its stations boasted service to an enormous range of destinations.

Throughout the middle part of twentieth century, passengers could transfer between bus stations in Hammond with relative ease—despite the fact that Greyhound and Trailways generally had separate stations. When one carrier relocated, the other often followed to a location within a few blocks of its rival—this happened *three times* in Hammond between 1950 and 1985. The companies, while fierce rivals, recognized that creating such synergy was to their mutual benefit.

In the early 1940s, the Greyhound Bus Station (4919 Hohman Ave.) and Union Bus Station (5036 Hohman Ave.) used by the various Trailways affiliates, were only a few hundred feet apart. Motor coaches destined from Chicago to Atlanta, Boston, New York, Washington, D.C., and many other points made stops in the community. Even some bus lines operating over routes from Chicago to Memphis, IN, New Orleans, and St. Louis, made stops in Hammond despite the fact that it was a considerable distance out of the way. These bus companies found the extra revenue generated to be worth the extra expense.

Hammond was also an important station on Great Lakes Greyhound's Detroit–St. Louis route in the late 1940s, which apparently has the distinction of being the only long-distance ground-transportation service ever operated through the metropolitan region without a stop in Chicago. By avoiding the circuitous trip into the greater Loop, these buses could travel on a more direct path between southwestern Michigan and central Illinois, shaving as much as 90 minutes from the schedule. To reach Chicago, travelers on these buses had to use connecting services from Hammond.

Greyhound moved its station to 52 State St., about a mile away from its rival's depot, in 1950. The new station, serving five bus lines, boasted 108 daily arrivals and departures in 1955, compared to the Trailways station's 43. Almost all of these arrivals or departures either originated or terminated in Chicago, but Gold Star Line's service between Hammond and Joliet, was an exception, another service that passed through the southern part of the region without entering Chicago.

A map from Greyhound's June 27, 1956 timetable shows Hammond as an intermediate stop for buses linking Chicago to New Orleans, St. Louis, and other points. (Author's collection)

Around 1960, the owner of the Greyhound station (which rented space to various other bus lines) allowed Trailways to move into its depot. Greyhound's concerns about the loss of business, however, soon led to this rival being told to move elsewhere. Trailways unsuccessfully filed suit to regain access to the station before opening its own depot—across the street—at 51 State St.[38]

By the late 1960s, the most prosperous years of the intercity bus industry had passed. Hammond still had 75 daily arrivals and departures in 1975, but the number was rapidly dropping. In 1980, the two carriers put their differences aside and moved into a relatively small, consolidated station near the Indiana Toll Road at 4040 Calumet Ave. The marriage, however, didn't last. A few years

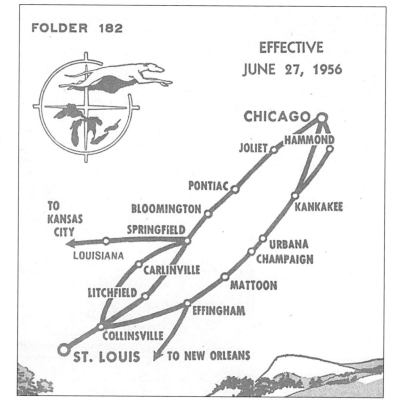

FOLDER 182

EFFECTIVE JUNE 27, 1956

CHICAGO
HAMMOND
JOLIET
KANKAKEE
PONTIAC
TO KANSAS CITY
BLOOMINGTON
SPRINGFIELD
LOUISIANA
URBANA CHAMPAIGN
CARLINVILLE
MATTOON
LITCHFIELD
EFFINGHAM
COLLINSVILLE
ST. LOUIS   TO NEW ORLEANS

HAMMOND'S GREYHOUND AND TRAILWAYS BUS STATIONS
51-52 S. State St., Hammond, IN

A variety of bus makes and models can be seen along Hohman Ave. in Hammond in the mid-1950s. The coach turning the corner, onto Sibley Blvd., appears to be a GMC (TDH 3714 model) transit bus, notable for having an upper row of windows for standees. (Richard Barnes collection)

A Greyhound bus travels through Hammond on September 22, 1963. This coach, operating express to Chicago, is likely near the carrier's station at 52 N. State St. (John LeBeau collection, courtesy of Mel Bernero)

later, Greyhound moved its station across the street to 4033 Calumet Ave., the third and final time the two carriers ended up with separate stations in Hammond less than a block apart.

Greyhound returned to the 4040 Calumet station in the early 1990s, but after Trailways withdrew from Hammond entirely, Greyhound and the affiliated Indian Trails Bus Lines decided to avoid the cost of a dedicated bus station altogether and instead moved their stops to a motel a few hundred feet north of the previous station near the Indiana Toll Road.

This, too, was only a stopgap solution. Like a stray, Greyhound then moved its service to an industrial building a few blocks farther north. The company eliminated the ticket agent from this station in 2011 and, along with Indian Trails, quit serving Hammond in the following year. Over a single generation, Hammond went from being a sizable intercity motor-coach hub to a city having no intercity bus service at all. The absence of service was soon filled by Miller Trailways, a regional operator, but Hammond presently sees only three pairs of buses making daily roundtrips, and none of its bus depots in service before 1990 survive.[39] ∎

A sign posted on the door at 3600 Calumet reminds anyone looking for a Greyhound bus that service from this location had ended. In the 1950s, Hammond had more than 80 daily intercity bus departures. (Author's photo)

Greyhound Replies to Bus Depot Suit

A headline appearing in the *Chicago Daily Tribune* on January 10, 1965, drew attention to a lawsuit filed by Trailways in an attempt to gain access to the Greyhound station in Hammond, IN.

In the process of downgrading Hammond service, Greyhound and Indian Trails moved into this nondescript industrial building at 3600 Calumet, an area devoid of sidewalks and other pedestrian amenities but near the I-90 interchange. Service ended in 2012. (Author's photo)

JOLIET UNION STATION
50 E. Jefferson St., Joliet, IL

Joliet Union Station is the only point in the metropolitan area outside of Chicago that has had long-distance passenger service on three different railroads in the past sixty years. Each of the station's service providers—the Chicago & Alton Railroad (Alton Route), Rock Island Lines, and Santa Fe—also ranked among the region's most prominent. Although this station's primary role was to accommodate passengers making trips starting or ending in the Joliet vicinity, for many years it was also a notable connecting point for passengers traveling between downstate Illinois and various points west.

This large Beaux-Arts structure, designed by Jarvis Hunt, was an instant landmark upon its completion

Joliet Union Station offers a striking backdrop for a pair of photographers on May 3, 1975. The graceful Beaux-Arts station was, by this time, oversized for the scale of its passenger operations. (Verne Brummel photo)

in 1912. With a stately entrance and tracks elevated during the station's construction, it has the aura of a big-city station. The Chicago & Alton and Santa Fe trains ran on parallel tracks through the station and crossed the Rock Island tracks only a few dozen feet from the station building. While this made it a fascinating location to watch trains, the track pattern was the source of considerable congestion for the railroads. Some trains stopping at the station blocked the crossing, creating a serious choke point for both freight and passenger trains.

JOLIET UNION STATION
50 E. Jefferson St., Joliet, IL

Passengers pass the time in Joliet Union Station on May 3, 1976. The doorway at the back of this upper-floor waiting room led directly to the platforms. Long-distance passenger trains still operated at the time over all three of the station's postwar passenger routes (those of the Rock Island Lines, Gulf Mobile & Ohio, and Santa Fe). (Verne Brummel photo)

The station had 58 long-distance trains in 1942, more than any other in the metropolitan region outside the city limits of Chicago. Passengers had the benefit of direct service to eight of the country's 20 largest cities—and seven of the 10 largest cities in the West. All of the famed Santa Fe *Chiefs* and Rock Island *Rockets* departing from Chicago stopped here, giving Joliet Union Station something that even the great depots in downtown Chicago lacked: competing service to Los Angeles and Phoenix. Twenty commuter trains also served the station at the time.

Despite being situated on what has been considered the dividing line between the country's eastern and western railroads, Joliet never became a significant hub for passenger connections between long-distance trains. The Alton Route's tracks between Chicago and St. Louis straddled that boundary, but none of the station's three carriers operated to points east of the Chicago dividing line. (The only trains operating to points east of Chicago made their last runs from Joliet in the early 1920s).[40] The result was a wedge-shaped pattern of routes radiating from the station, making most connections quite circuitous.

The network of long-distance routes serving Joliet Union Station, shown here as of January 1, 1949, was oriented heavily toward points south and west of Chicago. The Rock Island Lines' *Rockets* and Santa Fe Railway's *Chiefs* departed from adjoining platforms and offered competing services to Kansas City and Los Angeles. Passengers traveling between Bloomington, Springfield, and other downstate Illinois cities and the West often made transfers at Joliet between the Gulf, Mobile & Ohio Railroad's St. Louis–Chicago trains and these popular streamliners.

Still, Joliet Union Station was an attractive connecting point for passengers from Bloomington, Pontiac, Springfield, and other downstate Illinois points on the Alton Route (which was absorbed by the Gulf Mobile & Ohio in 1947) traveling to places west of Chicago, such as Denver, Des Moines, IA, and Omaha, NE. Moreover, Joliet's two main intercity bus depots—Greyhound Bus Depot (on the corner of Ottawa & Clinton) and Union Bus Terminal (5 W. Washington, often called "The Trailways Depot")—were just a few blocks away, making connections between buses and trains an attractive option.

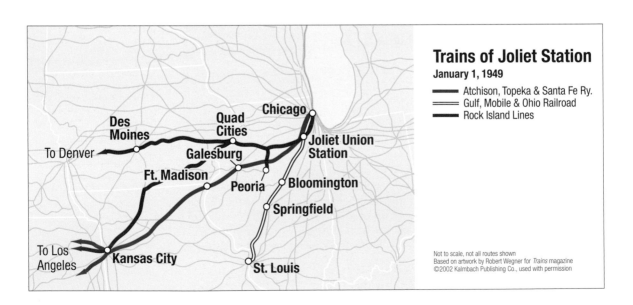

Trains of Joliet Station
January 1, 1949

— Atchison, Topeka & Santa Fe Ry.
= Gulf, Mobile & Ohio Railroad
— Rock Island Lines

Chicago
Quad Cities
Des Moines
To Denver
Galesburg
Ft. Madison
Peoria
Joliet Union Station
Bloomington
Springfield
To Los Angeles
Kansas City
St. Louis

Not to scale, not all routes shown
Based on artwork by Robert Wegner for *Trains* magazine
©2002 Kalmbach Publishing Co., used with permission

Rail service from Joliet suffered less from the postwar decline of long-distance train service than was the case with most stations, due to the strength of passenger business between Chicago and the West. The number of long-distance arrivals and departures remained a robust 46 in 1956. Transfers to intercity buses also remained a viable option after these services were consolidated in the mid-1950s into a new Union Bus Depot at 225 Chicago St., about six blocks away from Joliet Union Station.

The 1960s, however, spelled doom for many of Joliet's long-distance trains. Only 18 remained by early 1971, and after Amtrak commenced operations in May of that year, the number dwindled to 12. The count dropped to six when Rock Island (which did not join Amtrak) was finally allowed to exit the long-distance passenger business in 1978. Joliet's role diminished again when Amtrak's *Southwest Chief* was rerouted over the former Burlington Route main line through Naperville, in 1996, leaving Joliet with only commuter trains and Amtrak's Chicago–St. Louis trains on the former Alton–GM&O, ending its status as a connecting point for long-distance travelers.

By this time, however, a modest renaissance was under way. The number of commuters using Joliet Union Station gradually grew, and renovations to the station completed by the city government in the early 1990s restored much of its former grandeur. The upper-level waiting room was converted into a privately operated banquet hall, but the renovation provided passengers with a coffee shop and a more modern ticketing area, as well as improved lighting and climate control. In 2006, Amtrak added two additional round trips between Chicago and St. Louis. New commuter trains were added to the Heritage Corridor (formerly Alton–GM&O) route and along with those on the former Rock Island line using the station, attracted growing numbers of passengers.

Today, Joliet Union Station is a busy terminus for Metra trains on two routes from Chicago. Jarvis Hunt's striking landmark is today the only major station in the region used by long-distance trains during World War II that has not been demolished (or structurally modified) and remains in continuous use for transportation. Major modifications are nonetheless planned, in part, to eliminate the danger and inefficiency associated with having passengers boarding and disembarking Amtrak and Heritage Corridor trains walk across the former Santa Fe tracks. When this project is complete, the entire configuration of the station will change. Trains on the former Rock Island route will use a new platform farther east, allowing them to arrive and depart without crossing the "diamond" track junction. ∎

Led by a former Gulf, Mobile & Ohio locomotive, Amtrak Train 301, the *Prairie State*, exhibits the colorful mix of equipment typical of the new national carrier's first few years. Bound for St. Louis, the popular train calls at Joliet Union Station on August 22, 1971. For a few years in the early 1970s, this train and its running mate, the *Abraham Lincoln*, were extended beyond Chicago to Milwaukee, becoming the first scheduled passenger trains to operate through Chicago Union Station. (George H. Drury collection)

63RD ST.–WOODLAWN STATION
6327 S. Dorchester Ave.

An Illinois Central commuter train stops at 63rd Street Station while a CTA train for Jackson Park crosses the bridge in the distance. The canopy and platform used by long-distance trains is visible at the right edge of this November 14, 1966, photograph. (Ed DeRouin photo, courtesy of Bob Coolidge)

Dorchester Ave. and 63rd St. in Chicago is the only location in the metropolitan region where a traveler could at one time board all four of the region's primary forms of rail transportation—streetcars, electric interurban railways, the "L," and conventional railroads. In addition, a heavily used Greyhound station was only two blocks away. Such abundant transportation options created a wealth of timesaving connecting opportunities, not only for travelers destined for various neighborhoods in Chicago, but also for those traveling *through* the metropolitan region.

Illinois Central's depot on the 6300 block of South Dorchester Ave. was its primary stop in the Woodlawn neighborhood. This station, situated on "The Main Line of Mid-America," linking Chicago and New Orleans, was an important source of revenue in the 1880s and assumed added importance when it served throngs of fairgoers arriving for the 1893 World's Columbian Exposition. Some disembarked at 63rd St. and finished their journeys by foot or streetcar, while others rode trains that used specially built tracks that branched from the IC main line onto the fairgrounds.

As Woodlawn's prosperity continued in the early 1900s, the services available to railroad passengers improved greatly. In 1917, Illinois Central built a stately nine-story combination station and office building at 6327 South Dorchester. This new building housed the railroad's accounting department—supposedly located away from company headquarters to make it more difficult for executives to cook the books—and a restaurant and waiting room on the first floor. Passengers using the waiting room could reach the station platforms through an underground passageway.

The 63rd–Woodlawn Station boasted direct service to an enormous range of destinations. (A route map appears in the Central Station chapter on page 18). In the 1920s, more than 100 IC suburban trains stopped, as did several dozen long-distance trains, almost all of which originated or terminated at stations in downtown Chicago. The trains of the Big Four and Michigan Central (later assimilated into New York Central), including the *James Whitcomb Riley* and *Motor City Special*, also used these tracks and made stops here. South Shore Line trains began operating to the station in 1926, over IC's electrified commuter-service tracks that paralleled the main line.

Passengers arriving in the region who might otherwise have transferred between trains in downtown Chicago often saved themselves six or more miles of travel by reversing their direction at 63rd St. (Almost all trains using Central Station or the more northerly Randolph St. Station stopped at 63rd St.). For example, passengers traveling from Detroit on Michigan Central and connecting to New Orleans on IC could shorten their trip by transferring here.

An Illinois Central train headed for South Chicago is leaving 63rd St. Station circa 1935. The "intercity platform" from which this photo was taken was a stop for IC's *Panama Limited*, the Big Four (New York Central) *James Whitcomb Riley*, and other notable long-distance trains. An underground walkway linked this platform to the terminal building and office tower visible in the distance (center). (William C. Janssen photo, courtesy of the Shore Line Interurban Historical Society)

63RD STREET—WOODLAWN STATION
6327 S. Dorchester Ave.

A Metra Electric Highliner approaches 59th Street Station and another pair of trains can be seen in the distance at 63rd Street Station. Not a trace of the CTA "L" bridge once spanning the tracks is evident in this June 1994 photo. This was the last year for the Highliners to wear the IC orange paint scheme. (D.W. Davidson photo)

Another reason passengers detrained at 63rd St. was to connect to intercity buses and electric interurban railways. Greyhound's storefront station at 6302 Stony Island Blvd. was only two blocks away—a station that, for years, was the first stop after leaving downtown for *all* Greyhound and Indian Trails bus routes fanning out to the east and south. Connections to South Shore Line trains could be made without leaving the station, while passengers headed for Milwaukee had only a quarter-block walk to the 63rd–Dorchester "L" station, which, beginning in 1922, was the southern terminus of the North Shore Line interurban.

Although the North Shore Line stopped operating to the 63rd–Dorchester station in 1938, the other carriers maintained steady operations through the busy wartime years. However, the range of services available gradually dwindled after hostilities ended. Streetcar service ended in 1953. Michigan Central trains ceased using the IC route in 1957.[41] Greyhound relocated the last of its South Side stops to 95th St. shortly after the Dan Ryan Expressway's rapid-transit line opened in 1969—one of several changes governed by a deal with the city to create new stops near major rapid-transit hubs. Long-distance passenger trains made their last stops at 63rd St. with Amtrak's debut in 1971.

The station remained a stop for both local and express suburban trains after IC introduced zone schedules in 1974, but this was short-lived. In 1975, the express trains began stopping at 59th St. instead. South Shore Line trains also continued stopping, to serve students at the nearby Mount Carmel High School, but the number doing so gradually thinned (reportedly in part due to the problem of conductors getting robbed on the platform). Even the "L" was on borrowed time; in 1982, inspectors declared the bridge over the IC tracks to be unsafe, resulting in the eventual elimination of the route through the entire neighborhood.

Visitors today may find it hard to imagine that 63rd and Woodlawn, which is surrounded by a considerable amount of vacant land, was such a busy and diversified transportation hub 60 years ago. The former IC office tower was razed in the late 1980s. Other buildings were cleared away for the Apostolic Church of God, a large, predominately African American place of worship, whose leaders had pushed for the removal of the L structure. Metra trains still stop at the surviving portion of the old railroad station, but 63rd St. today is only a minor Metra stop, bypassed by express trains.

Still, whispers of the past can be heard when passengers must make transfers between mainline and South Chicago trains at the two surviving platforms. These modest wooden structures, and remnants of the beams and passageway part of the station and office building, are all that remains of a connecting complex that catered to local and long-distance travel for more than 70 years. ∎

A southbound Metra train to University Park, equipped with new Nippon Sharyo cars, speeds past 63rd Street Station in August 2013. Just out of view, beyond the platform on the right, are remnants of the portal leading to the "intercity platform" last used in 1971. (Xhoana Ahmeti photo)

OUTLYING BUS AND RAIL TERMINALS
Terminal Town

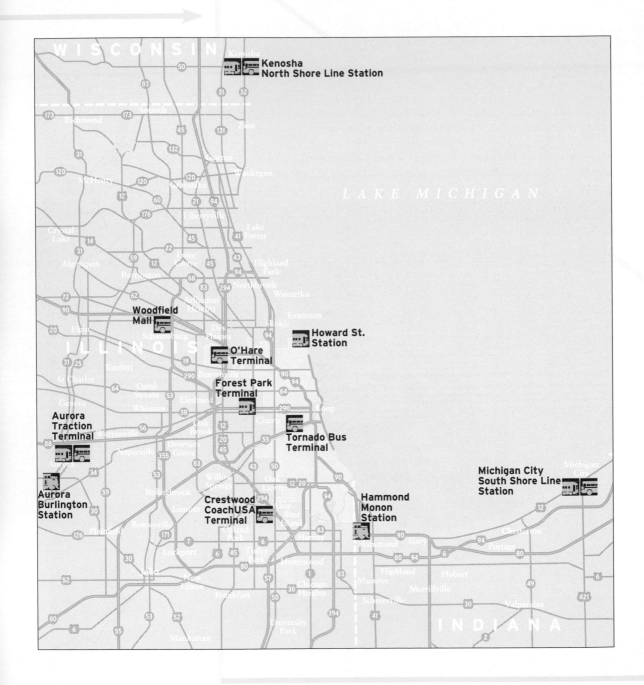

Relatively few passengers traveling by train or bus from points outside of metropolitan Chicago since 1939 have made transfers at an outlying location before reaching downtown. Almost all passenger trains and intercity buses destined for the region have terminated near the Loop.

In this respect, Chicago's transportation system has been remarkably passenger-friendly compared to those of metropolitan New York, Los Angeles, and San Francisco. In those regions, a far greater share of buses and trains have ended their runs a considerable distance from the central business district. Even today, many Manhattan-bound passengers make transfers daily at railroad stations in New Jersey. Los Angeles Union Station is far enough removed from the central business district that most commuters must transfer between trains or buses to reach downtown offices. San Francisco's long-distance trains terminate on the east side of its namesake bay rather than in the city proper.

Between the late 1920s and the early 1930s, all regularly scheduled passenger trains operating over conventional railroads into the region, except for a few Burlington Route trains that ended their run in Aurora, IL, terminated in downtown Chicago. As the economics of passenger service worsened, the Wabash Railway in 1931 dropped service between Dearborn Station and Toledo, and in its place began operating a mixed train (one carrying both freight and passengers) between Montpelier, OH, and Gary. In 1933, the Chesapeake & Ohio of Indiana began terminating an inbound train at Hammond, IN.

These trains, while important to some travelers, barely warrant footnote status in the region's train-travel history. By 1953, all of them were gone, and all of Chicago's passenger trains operated all the way downtown. Nor did railroads stop using the terminals in the tumultuous years leading up to Amtrak, when some were apparently eager to dissuade passengers from riding. Circumstances were different on bus lines and the electric interurban railways. As described in this section, several stations were necessary transfer points for passengers traveling to downtown Chicago from points outside the metropolitan region. Several bus routes ended their runs at outlying stations on these electric railways.

Today, all of the historical feeder routes are gone. Almost all bus and train travelers from outside the region heading for downtown Chicago are offered one-seat rides. In a region as complex as Chicago, however, it is inevitable that there are exceptions. The Coach USA buses operating to suburban Crestwood, intercity buses serving O'Hare Airport, and campus-oriented bus lines operating to Schaumburg, IL, all terminate at outlying points. Similarly, transit buses from Kankakee, IL now shuttle passengers to the Metra station in University Park (see Appendix II for a more complete list of locations). None of these services, however, is significant in true long-distance travel to or from the Loop.[42] ■

(Opposite) This map shows the eight locations outside of downtown Chicago that have been terminal points for intercity trains and buses since 1939. Many travelers arriving at these stations needed to make transfers to reach downtown. The interurban railway stations in Aurora, Kenosha, Forest Park, and Michigan City were endpoints of either feeder-bus routes or trains. Conventional passenger trains ended their runs at Aurora's Burlington Route Depot and Hammond's Monon Station.

AURORA "BURLINGTON ROUTE" DEPOT
175 S. Broadway St. & 11ᵗʰ St., Aurora, IL

The vista-domed *California Zephyr* is 38 miles into its westbound run at Aurora's Burlington Route station, circa 1965. The two platforms at the far right were used by Chicago-bound suburban (commuter) trains, while the longer platform at left primarily served long-distance trains. The head house, set at an angle to the tracks, is visible in the distance. (Dave Weber photo, courtesy of Michael Spoor)

The Chicago, Burlington & Quincy Railroad (Burlington Route) depot in Aurora was a place of many "lasts": it was the last of the Chicago region's outlying stations to serve as the terminus for an intercity train; it was the last of the region's historic stations to cease being a stop for Amtrak trains; and it was last of the stations and terminals featured in this volume to be completely demolished. As a stop and transfer point for some of Chicago's most celebrated passenger trains years ago, however, this station ranks high in historical importance among the region's outlying stations.

Like so many of its contemporaries, this flat-roofed, two-story, brick station, opened in 1922, had distinctive Beaux-Arts features, a style still in favor at the time. Aurora's lengthy platforms were protected under umbrella canopies and linked by an underground pedestrian tunnel, giving the station a big-city feel. With a restaurant, spacious waiting room, and upper-floor offices, this building's size was a salient reminder of the Burlington Route's dominance in this part of the region.

The station was almost constantly active during daylight hours, accommodating passengers traveling on five Burlington Route lines fanning out from this community. Being the endpoint of the famous "racetrack" from Chicago to Aurora—a busy commuter and long-distance passenger route with three tracks—also gave it a high level of visibility among travelers. Less than a mile west of the station, the busy main line split into two routes, with one branching to Kansas City (route of the *American Royal Zephyr*) and Denver (*California Zephyr* and *Denver Zephyr*) and the other to the Twin Cities (*Empire Builder*). Less used lines also linked Aurora and vicinity to Streator and West Chicago.

The Burlington Route's *Empire Builder* and all its famed *Zephyrs* operating to and from Chicago stopped in Aurora. For most, it was their first westbound stop and last eastbound stop, thereby attracting travelers from a wide geographic region. More than two dozen weekday commuter trains also covered the 39-mile distance to and from Chicago Union Station. Unlike the similarly sized Union Stations in Gary and Joliet, Aurora was a "terminal" in the strict sense of the word. Trains from Rockford, Streator, and West Chicago, and Davenport, IA, all terminated here in the 1920s. The six-track station was an important transfer point for passengers traveling to and from Chicago and branch-line points.

Some of the luster of this station faded as the economics of passenger-train operations deteriorated, with the trains to West Chicago and other

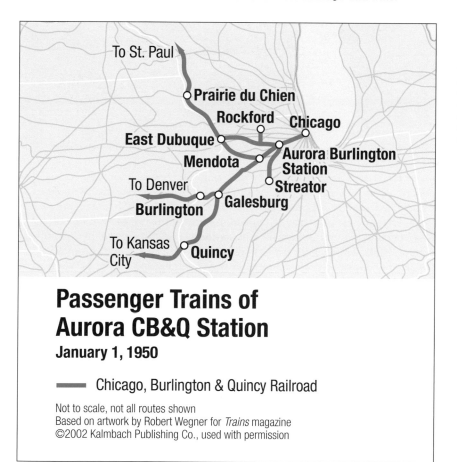

Passenger-train lines emanated from the Aurora Burlington Route station in many directions on July 1, 1950. Most of the famed *Zephyrs* en route to the West made their first outbound stop at this six-track depot. Branch line trains from Streator, IL, terminated at the station, as did those from Rockford in earlier years, requiring passengers destined for downtown Chicago to make connections.

Passenger Trains of Aurora CB&Q Station

January 1, 1950

—— Chicago, Burlington & Quincy Railroad

Not to scale, not all routes shown
Based on artwork by Robert Wegner for *Trains* magazine
©2002 Kalmbach Publishing Co., used with permission

AURORA "BURLINGTON ROUTE" DEPOT
175 S. Broadway St. & 11th St., Aurora, IL

Built in the preferred architectural style of the Chicago, Burlington & Quincy Railroad, the Aurora station faced South Broadway. Its six station tracks were on a high embankment (visible at right). An enclosed walkway below the platforms linked the waiting room with the pedestrian tunnel that protected passengers from inclement weather and allowed them to avoid crossing active tracks. The station's hemmed-in location, however, limited the parking available to commuters. (Verne Brummel)

intrastate points being among the first to go. The Rockford service, reduced to self-propelled gas-electric cars, ended in 1949. But the similarly downgraded Aurora–Streator service continued to run, becoming the last intercity passenger train to operate in metropolitan Chicago without using a terminal in downtown Chicago. Passengers on these trains and those on long-distance trains often transferred in Aurora to catch commuter trains making local stops between Aurora and the Loop. Although the Streator trains stopped running in 1953, the Burlington Route's commuter business grew after the Chicago, Aurora & Elgin Railroad ceased operating in 1957. As late as 1962, 21 daily intercity trains and 24 commuter trains used the station.

By the late 1960s, however, even the most celebrated *Zephyrs* seemed headed for oblivion. The *California Zephyr* had made its last run in 1970, and on Amtrak's debut the following year, the *Empire Builder* stopped calling at Aurora, having been rerouted via Milwaukee. Intercity service at Aurora dwindled to two daily round trips, the Chicago–Oakland *San Francisco Zephyr* and the *Illinois Zephyr*, a train from Chicago to West Quincy, IA.

The Aurora Burlington Station's spacious waiting room offered a pleasant setting for passengers. By the time this photograph was taken in July 1974, the station was in the hands of Burlington Northern and responsibility for its long-distance trains had been turned over to Amtrak. A careful inspection of this photograph reveals the chandeliers, classic wooden benches, storage lockers, and abundant natural light from the large windows. The station restaurant, however, had already closed. (Verne Brummel)

The cost of maintaining the building, coupled with the declining fortunes of downtown Aurora and chronic shortages of parking on the property, made it a target for elimination. The end came on December 7, 1986, when the commuter-train terminus was moved to the Aurora Transportation Center, a new multi-modal facility about a mile away. Amtrak relocated its area stop to neighboring Naperville, giving Aurora the dubious distinction of being one of the few communities with a population of more than 75,000 that once had Amtrak service, but came to find that carrier's trains whisk through town without scheduled stops. The old depot was finally demolished (see page 258) in 2013. ■

This classic railroad scene calls to mind Aurora's role as a busy passenger hub. A passenger walks down the middle platform toward the Burlington Route commuter train at the Aurora Station in 1964. The passengers in the distance have apparently disembarked a long-distance train and are walking toward the pedestrian tunnel that will lead them to the station building. (Thomas E. Almond photo, courtesy of John Almond)

AURORA TRACTION TERMINAL
2 N. Broadway St., Aurora, IL

Once an important transfer point, Aurora's Traction Terminal Building sits vacant in April 2012, more than 70 years after it last served a terminal function. The Chicago, Aurora & Elgin Railroad had its waiting room and ticket office on the ground floor. Trains operated down the middle of Broadway St., which runs from left to right in this photo along the narrow side of the building. (Author's photo)

The Traction Terminal at 2 N. Broadway St. in Aurora was once a hub for a streetcar line, three electric railways, and several intercity bus lines. In addition to serving Aurora's large local population, this station and its eventual replacement were important bus-to-train transfer points for passengers destined for Chicago.

The six-story Hotel Arthur was one of Aurora's tallest buildings when it opened in 1905. The Aurora, Elgin & Chicago Railroad leased the building in 1915, renaming it the Traction Terminal Building and transforming the ground floor into a waiting room and diner. The AE&C's company offices were moved to the upper floors. Trains began departing from the middle of Broadway St. directly in front of the building.

With the trains of three companies—the AE&C; the Aurora, Plainfield & Joliet; and the Chicago, Aurora & DeKalb—the terminal was a hive of passenger activity. The era of train-to-train transfers, however, was relatively short-lived. The interurban railways to DeKalb and Joliet ceased operation in 1923 and 1924, respectively. Suffering from a downturn in business, the AE&C was reorganized as the Chicago, Aurora & Elgin, which opted not to renew its lease of the Traction Terminal. The company, instead, moved in 1935 into a storefront location a few hundred feet north, on the southwest corner of Fox St. (now East Downer Pl.) and Broadway.

Four years later, following a major project that moved its tracks off Broadway, the company built a new route in downtown Aurora on a private right-of-way hugging the east bank of the Fox River and including a new riverside station equipped with a rapid-transit-style, high-level platform. Passengers could reach the platform directly through the station's back door. The ticket office was moved to a storefront location at 56 N. Broadway, near the intersection of Broadway and New York Ave. This facility also served as the Aurora stop for three intercity bus lines: Burlington Trailways, Joliet–Aurora Transit Lines, and Greyhound.[43]

The bus lines funneled a great deal of passenger business to the CA&E over the following 18 years. Aurora was the first stop made by many of Greyhound's transcontinental buses operating west from Chicago, offering direct service to Los Angeles, San Francisco, and Seattle. In 1947, the station had 23 daily intercity bus arrivals and departures.[44]

Four daily Burlington Trailways buses originating in Somonauk, IL, a small town 22 miles southwest of Aurora, fed a large amount of traffic to the interurban, picking up passengers in Oswego, Plano, Sandwich, and Yorkville, IL, before terminating at the Aurora depot. Passengers arriving on these coaches had only a short walk to board Chicago-bound CA&E trains. Bus travelers also disembarked here to reach suburban destinations, including Lombard, Warrenville, and Wheaton, IL, along the interurban railway.

This colorful era of bus-train transfers lasted until the CA&E made its last run in July 1957. Without the trains, the Trailways buses from Somonauk had little purpose and vanished from timetables that same year. Gradually, the Aurora location assumed the character of a traditional intercity bus stop, almost exclusively serving passengers originating or terminating their trips in Aurora with few connections.

Much as the Traction Terminal had been abandoned for transportation a half century before, the intercity bus companies vacated Aurora's North Broadway St. station (by this time called Union Bus Depot) around 1974 and moved into a smaller station at 13850 River St. By 1980, the number of bus departures had diminished to seven. In 1995, all scheduled long-distance bus operations were moved to the Aurora Transportation Center at 233 N. Broadway, a facility also used by Metra trains. Greyhound's use of the center lasted only until 2011, when it eliminated Aurora from a shrinking network of routes and left the community without any intercity bus (or rail) service.

The Traction Terminal (now often called the Terminal Building) survives as a looming reminder of Aurora's importance as a connecting hub. The building, added to the National Register of Historic Places in 2005, is now vacant, and virtually all other traces of the CA&E, including the 1939 station built along the river, are gone. Efforts to preserve the old high-level platform that had— remarkably—eluded demolition until the early 1990s ended unsuccessfully. ■

Signage above the doorway at 2 N. Broadway in Aurora offers a reminder of the structure's heritage. (Author's photo)

A two-car CA&E train stands at the end of the line in Aurora, on April 24, 1957. This station, replacing the Traction Terminal, had a high-level platform linked to the waiting room by the walkway visible at lower right. The rear of the bus station can be seen in the distance (right center), above the open portion of the platform. (Paul Stringham photo, William Raia collection)

CRESTWOOD COACH USA TERMINAL
5545 W. 127th St., Crestwood, IL

The Coach USA Terminal in south-suburban Crestwood has been a hub for passengers traveling to and from Chicago airports for more than 25 years. Service from this tiny facility has rarely stayed the same for an extended period, but at one time it encompassed trips to more than 20 destinations in three states.

Interest in express motor-coach service to O'Hare grew sharply after the opening of the Tri-State Tollway in 1958. Buses that were once forced to use slow arterial highways could now move at speeds exceeding a mile a minute, putting a vast geographic area little more than an hour's drive from O'Hare.

Advertisements emblazoned on a Coach USA bus and a billboard tout the convenience of the carrier's services to Midway and O'Hare airports. The Crestwood Terminal, built by Tri-State Coach Lines, is equipped with a small waiting room and ticket counter and has remained largely unchanged since its 1988 opening. (Author's photo)

In December 1964, Tri-State Coach Lines launched air-conditioned limousine service to the airport from the Hotel Gary (in downtown Gary) and the Hammond Holiday Inn along Interstate 80 near the eastern end of the Tri-State Tollway. Seven vehicles operated in each direction (fewer on Saturdays), making the 60-mile trip from Gary to O'Hare in an hour and 15 minutes.

This privately operated service performed well, allowing service to be gradually expanded. Intercity bus service, however, was still tightly regulated by state and federal agencies at the time, and the carrier soon found itself in a regulatory showdown with Continental Air Transport, which according to the *Chicago Tribune* had "nearly identical plans" to expand to the south suburbs and Northwest Indiana.

The Illinois Commerce Commission sided in favor of Tri-State, and by 1969 the company was operating hourly from Gary to O'Hare. Business was strong enough that the company gradually replaced its limousines with buses. By 1970, an extensive network of routes south of O'Hare was in place, with services terminating at Park Forest (a booming suburb at the southern edge of the metropolitan region), Harvey (an older industrial suburb south of Chicago), and South Bend, IN. The carrier also picked up many passengers at intermediate stops in Chicago Heights and Homewood, IL; Portage, and Merrillville, IN; and other locations in Illinois and Indiana.

Such robust expansion, however, did not come without problems. Many buses and limousines now operated up and down the Tri-State Tollway only a few minutes apart. Some buses may well have operated with few passengers during off-peak hours. To achieve greater efficiency, the company created a small hub in south-suburban Alsip, IL, near the Tollway's Cicero Ave. interchange. It began consolidating passengers originating at different locations en route to O'Hare at this new hub, which first appeared in timetables in 1972.

The hub was initially located in the parking lot of the Alsip Holiday Inn, with the hotel's lobby serving as the waiting room. Changes were almost constantly made in response to customer demand. For example, the Park Forest route was dropped, Michigan City and Valparaiso, IN, were added, and a route to South Bend was turned over to an affiliate, United Limo. Through it all, Alsip remained the focal point of operations throughout the south suburbs.

The company's expansion reached new heights when it began serving Kankakee, IL, in 1980 and New Buffalo and Stevensville, MI, in 1984. The introduction of service to resurgent Midway Airport several years later further energized the Alsip hub. Buses destined for Midway and O'Hare now arrived at Alsip almost simultaneously. Passengers were given a few minutes to switch buses to reach their airport of choice while the drivers transferred baggage.

The scale of this operation grew so sharply that in July 1988 the company moved the hub from the hotel to a newly built terminal about a mile west in Crestwood. Although the short distance added a few minutes to travel times, the new facility had a small waiting room, a ticket counter, and a circular driveway to allow for efficient bus turnarounds. It also had a large parking lot, prominent signage, and a large apron for bus loading and unloading.

The hourly cycle of departures to Midway and O'Hare is under way at the Crestwood terminal on June 3, 2012. The front coach is bound for O'Hare, and its Midway counterpart is second in line. (Author's photo)

Crestwood's hub role continued after Coach USA acquired United Limo and Tri-State Coach in 2002, but the company gradually eliminated smaller stops in favor of larger stations. Within a few years, service to southern Michigan and Kankakee, Matteson, and Merrillville had all ended. By 2010, all that was left were three high-intensity routes linking Crestwood with Midway, O'Hare, and east to Highland, IN, Michigan City, and South Bend. On these routes, however, heavy traffic often necessitates that the company operate several buses to meet the demand.

Today, the Crestwood terminal is a busy place, with 70 daily bus departures, and still serves as an important transfer point, but lacks service to points outside the Chicago metropolitan area besides South Bend. ∎

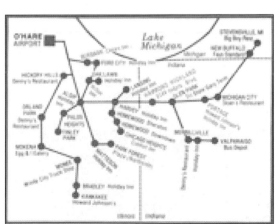

The Tri-State Coach network, encompassing bus service to 30 stations, is depicted at its maximum size in this April 1985 timetable map. Routes radiated from the Alsip hub (later replaced by today's Crestwood hub) in five directions and stretched as far as Stevensville, MI, at the time. (Author's collection)

A Chicago, Aurora & Elgin Railroad train is at the Forest Park Station in March 1954. Passengers headed for downtown Chicago on arriving CA&E trains had to make transfers here to CTA trains from 1953 until 1957, which was about the only interval during the 1950–62 period that this station was not undergoing change. The train that appears here is notable to railroad historians by showing how the carrier often cobbled together different types of equipment, even as it cut back service to Forest Park. (Krambles-Peterson Archive)

Forest Park's role in intercity travel evolved in ways much different than in most inner-ring suburbs. Cutbacks rather than the expansion of service twice elevated its status, first by making it a transfer point for downtown-bound passengers on an electric interurban railway, the Chicago, Aurora & Elgin Railroad (CA&E), and later by making it an important stop on several Greyhound routes. In its most recent incarnation, Forest Park is back to being a hub exclusively supporting local transit providers.

When the Aurora, Elgin & Chicago Railroad (which later became the CA&E) began service between Aurora and Forest Park in 1905, it established its primary local station on Desplaines Ave. (later Des Plaines Ave.). East of this station, trains proceeded to Laramie Ave. on Chicago's West Side, where there was a connection with the Metropolitan Elevated ("Met L"). Trains using this route almost immediately became the preferred method for traveling between Aurora, Elgin, other outlying communities, and downtown Chicago. After reaching downtown over the "L," trains terminated at Wells St. Station (described elsewhere in this volume). The Met L (which later became part of the CTA) also began operating beyond Laramie Ave. to Forest Park.

Forest Park's role as regional hub continued to grow. The Met L began running over a new branch line opened from Forest Park to Bellwood and Roosevelt Rd. in 1926.[45] Extensive bus service was provided from the station as well. The CA&E and "L" had separate platforms on opposite sides of Des Plaines, but operating practices appear to have varied over the years.

Despite Forest Park station's expanding range of services, it was still not a significant transfer point for passengers making long distance trips. The Chicago Great Western and Soo Line depots in Forest Park were less than a block away, but these companies operated only a handful of daily trains and no doubt generated little transfer traffic.[46] Almost all travelers changing trains were on short-distance trips within the city and close-in suburbs.

The construction of the Congress Expressway in the early 1950s dramatically changed operating practices. Building the superhighway required demolition of the Garfield "L" (part of the old Met L), forcing trains to be moved to a new "temporary"

FOREST PARK C&AE/CTA STATION
711 Des Plaines Ave., Forest Park, IL

terminal in Forest Park. In the midst of this project, in September 1953, CA&E discontinued service to downtown Chicago, thus making it necessary for all passengers headed there to transfer to "L" trains in Forest Park. Suddenly, passengers traveling between a wide range of communities, including some outside of the metropolitan region (see the chapter on the Aurora Traction Terminal earlier in this section), made connections at the Desplaines Ave. station.

This cumbersome arrangement lasted four years, until CA&E ceased operating entirely on September 20, 1957. Two years later, Forest Park station was replaced by another new "temporary" facility on the same parcel as the old one. This new station had an extra platform in case the CA&E resumed service, but this never occurred.

After this second temporary station had been in use for more than 20 years, largely due to a chronic shortfall of funding, it was replaced in 1980 with a much larger station that remains in use today. Measuring 432 feet long and 80 feet wide, the new station's large canopy completely covered the island platform, making it a prototype for future rapid-transit superstations. Greyhound relocated its Oak Park stop to this spacious facility that same year, thereby making Forest Park once again an important transfer point on intercity trips. In 1980, 11 daily Greyhound buses, some destined for the West Coast, made stops. Passengers between a few cities could save time by transferring here rather than in downtown Chicago, but the station mostly served passengers starting or ending their trips in the area.

Greyhound's presence lasted approximately 13 years.[47] Since it dropped its Forest Park stop in 1993, the station has served only as a connecting point for passengers moving within the metropolitan region. ∎

HAMMOND MONON STATION
Douglas St. and Lyman Ave., Hammond, IN

The Monon Route's tidy Hammond Station was the last depot built in the Chicago region that served more than one railroad—and one of the first to close due to passenger-train cutbacks. Among its most memorable feature was the modernist sign advertising the Monon and Erie Railroads. (Donald Kaegebein photo)

The Monon and Erie Stations in Hammond, if treated as a single depot by virtue of their side-by-side location, was one of only two points in the metropolitan region outside of the city limits of Chicago served by at least three passenger railroads in the postwar era. (The other was Joliet Union Station, featured on page 142). The trains of the Chesapeake & Ohio of Indiana (C&O), Chicago, Indianapolis & Louisville (known as the Monon), and Erie railroads stopped here.

The original stately and neoclassical Monon depot also holds the distinction of being one of only two stations outside of Chicago (with the other being the Burlington Station, in Aurora, discussed earlier in this section) to be the terminus of a bona fide long-distance passenger train in the postwar era. This station became such a terminus in June 1933, when C&O replaced its overnight run linking Cincinnati and Chicago's Central Station with a day train ending its trip here. Passengers seeking to reach downtown Chicago, consequently, had to make connections with the station's other two carriers, both of which honored the C&O's tickets. Some simply crossed the tracks to the adjacent Erie Station, a two-story wooden structure, and rode that carrier's train to the city.

The track configuration resembled an upside-down Y. All southbound and eastbound trains reached these stations over the stem—the Chicago & Western Indiana Railroad—originating at Chicago's Dearborn Station. The routes split just north of the Hammond station area, with Erie and C&O trains veering southeast to Griffith, IN, and those of the Monon following a more southerly trajectory toward the town in Indiana after which it was named.

This station complex was a vibrant place during World War II. In 1942, the depots served 16 daily passenger trains—eight on the Erie, six on Monon, and the daily C&O pair. Monon's *Tippecanoe* and *Hoosier* linked Chicago and Indianapolis, while *The Thoroughbred* connected Chicago and Louisville. Erie's best-known train of this era, the *Lake Cities,* departed daily for the Windy City and Jersey City (later to Hoboken), NJ.

Although C&O discontinued its little-used train to Cincinnati in 1949, the rest of this pattern of arrivals and departures remained much the same after the war. A sharp drop in traffic in the early 1950s, however, necessitated that station expenses be cut. The two stations were replaced on October 27, 1953 with a consolidated, but much smaller, Monon Station a short distance south, which consisted of little more than a waiting room, ticket counter, and a modest parking lot. New rail-passenger station openings were so rare at the time that a *Chicago Tribune* journalist declared, "Look! Somebody's building a new railroad passenger station."[48] The Monon also built new stations in Lowell and Monon, IN, to shore up its image, and celebrated their opening by operating a special train for company officials and dignitaries.

At first, Hammond's replacement station appeared to have a promising future, but its role diminished when Monon's Indianapolis trains made their last runs in 1959. The *Thoroughbred* stopped running in 1967, leaving only Erie Lackawanna's *Lake Cities*. (Erie having merged with the Delaware, Lackawanna & Western in October 1960 to become the Erie Lackawanna Railroad). The merged company dropped this, its last remaining intercity

The ornate Hammond Monon Station, notable for its decorative windows, was demolished as part of the transition to the much smaller 1953 station. (Donald Kaegebein photo)

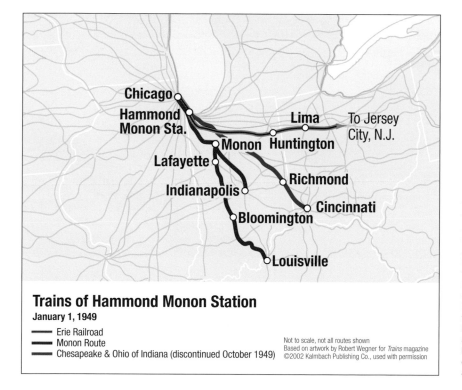

Trains of Hammond Monon Station
January 1, 1949

—— Erie Railroad
—— Monon Route
—— Chesapeake & Ohio of Indiana (discontinued October 1949)

Not to scale, not all routes shown
Based on artwork by Robert Wegner for *Trains* magazine
©2002 Kalmbach Publishing Co., used with permission

The passenger-train routes serving the Monon station in Hammond, IN, as of January 1, 1949, offered direct service to Cincinnati, Louisville, and Jersey City, NJ. These trains, operated by the Chesapeake & Ohio of Indiana, the Monon Route, and the Erie Railroad, respectively, also provided convenient access to more than 40 Hoosier State communities. The C&O train from Cincinnati ended its run here, but both the Erie and Monon as well as the South Shore Line, which had a station nearby, honored that carrier's tickets between Hammond and Chicago.

HAMMOND MONON STATION
Douglas St. and Lyman Ave., Hammond, IN

passenger train, on January 6, 1970. Hammond's replacement station was boarded up and then demolished in the 1980s. A regional transportation agency acquired the former Monon tracks and right-of-way for possible commuter service. Hopes remained high that passenger trains might one day return. This abandoned right-of-way was instrumental to a proposed commuter service linking Chicago and rapidly growing suburbs in Lake County, IN. As optimism waned in 2012, however, the rails and crossties were removed to allow for the creation of a bicycle path.

Steam for train heating escapes from the lead locomotive of Erie Lackawanna's Chicago–Hoboken, NJ, *Lake Cities* at Hammond on March 31, 1964. The platform is a considerable distance from the Monon Route platform next to the station building (visible at far left). The two companies' lines diverged just south of the depot. (Marty Bernard photo)

Visitors to the site today will find nothing more than concrete platforms, crumbling with age, and an earthen fill that once supported a tidy station that no longer exists. The prospect of commuter trains stopping here remains a distinct possibility, but the funding issues are proving difficult to resolve. This service, if instituted, would use the South Shore Line between Chicago and Hammond, and the former Monon route beyond to St. John, IN. ■

HOWARD ST. NORTH SHORE LINE/CTA STATION
7519 N. Paulina St.

Heading south on Track 3, a North Shore Line *Electroliner* train from Milwaukee (left) stops at Howard St. in January 1963, a few days before the end of North Shore operations. North Shore trains drew power from an electrified third rail over most of the CTA system, and from overhead wires when on their own track. The trolley poles used to reach overhead wires are lowered on this train because there is a third rail on this CTA track. (Ed DeRouin photo, from the collection of Lou Gerard)

Howard St. was for many years an important transfer point for trains operated by the Chicago Transit Authority (CTA) and the North Shore Line interurban railway. Greyhound Lines also operated its motor coaches from a storefront station at this busy junction. Although the station is no longer an intercity-travel connecting point, it remains a vital transfer point for passengers using CTA trains and buses.

Travelers today primarily associate Howard St. Station with the elevated rapid-transit service that has been operating to this area since 1908. In 1919, the Chicago Elevated Railway began terminating some trains here rather than in Evanston to allow for more efficient operations, thereby beginning the custom of having Evanston-bound travelers transfer at Howard St. That same year, the North Shore Line began operating trains over the "L" between Howard and downtown Chicago, a lucrative route between Milwaukee and the Windy City.

This station's role as a transfer point grew sharply after the opening of the last portions of the North Shore Line's "Skokie Valley Route" between Howard St. and Waukegan, in 1926. This new high-speed route, branching off from the older shore line route at Howard St., was a faster and less-urbanized corridor to Waukegan that shaved 20 minutes from the trip to Milwaukee. As more trains operating to and from Milwaukee began using the new route, more passengers starting their trip on the North Shore's original route near the shoreline made transfers at Howard St.

The North Shore Line and "L" trains shared the route through Howard St. station for many years, but the carriers operated from separate platforms and maintained separate ticket offices at the station. The CTA (created in 1947 to take over bankrupt rapid-transit lines) also grew more complex over time. The transit company ran trains over the North Shore's Skokie Valley Route as far as Niles Center in Skokie—which is today the Yellow Line—while also acquiring the interurban railway's shoreline route as far north as Wilmette.

By the late 1940s, Howard St. was also an important stop for intercity buses. Northland Greyhound had a station near the station's front door, at Little Al's Record Shop at 329 Howard, which also had a four-seat waiting area and a single bathroom for men and women.[49] Thirty of its buses stopped there daily in 1946, carrying passengers to Milwaukee in competition with the North Shore Line, as well as providing direct service to Minneapolis–St. Paul. United Motor Coach, meanwhile, operated frequent buses from Barrington, IL, to Howard St. over a route that was still heavily rural.

Large numbers of local and intercity travelers continued to make transfers at Howard into the early 1950s, but the neighborhood struggled with "white flight" and the loss of business to the outlying suburbs. Greyhound gradually reduced its service and petitioned in 1952 to move its Howard St. stop to

A northbound North Shore Line train operating over the Skokie Valley Route is on Track 4 at Howard St. station in June 1962. This view looks east onto Howard St. Today, this station is used by CTA Red, Yellow, and Purple Line trains operating between Chicago, Wilmette, and Skokie. (Donald Duke, courtesy of the Shore Line Historical Society)

HOWARD ST. NORTH SHORE LINE/CTA STATION
7519 N. Paulina St.

A two-car Howard–Jackson Park CTA train negotiates a switch at Howard St. Station in 1979. The North Shore Line's Skokie Valley Route, having been transformed into the Skokie Swift (today's Yellow Line), can be seen slipping below the Chicago Ave. overpass at upper left. Note that the platforms are still made of wood. (Mark Llanuza photo)

Davis St. in downtown Evanston. Although many riders supported the move, it never occurred.

By the end of the decade, it was evident that both the North Shore and intercity bus services from Howard St. were at risk of elimination. The opening of the Kennedy Expressway in 1960 created a continuous expressway route between downtown Chicago and the northern suburbs by way of the Edens Expressway (portions of which had been completed as early as 1951). The North Shore Line's ticket sales gradually fell, and federal authorities allowed it to end operations on January 21, 1963. Greyhound service dwindled to three daily buses in each direction before it moved its terminal from Howard St. to Skokie's North Shore Station in the mid-1960s.

Greyhound's exit left only the CTA and suburban bus lines at Howard St. Station, so it reverted to being a connecting point solely for transit riders. Most passengers today associate Howard St. with being an important point for transfers between the Red, Purple, and Yellow Line CTA trains. Relatively few are aware of its once-significant role in interstate travel by bus and train up the Lake Michigan shoreline. ■

This map appeared in a 1926 North Shore Line advertisement that described the new Skokie Valley Route as one that "typifies the last word in transportation." This route, notable for being fast and straight, shaved more than 20 minutes from the Chicago–Milwaukee trip. (J.J. Sedelmaier Productions Collection, courtesy of Shore Line Interurban Historical Society)

Service Starts
JUNE 5th
on the Skokie Valley Route

NORTH SHORE LINE

The high-speed electrically-operated railroad

ON Saturday morning, June 5, the Skokie Valley Route of the North Shore Line will be opened with full service. This new twenty-three miles of double track will prove of benefit, directly and indirectly, to every community served by the North Shore Line.

Increased Service on the Shore Line Route

First of all, by relieving traffic on our Shore Line Route, the new line will enable us to render increased service to all North Shore suburban communities. Such increased service will be started immediately.

Commencing June 5, three types of town-to-town service will be supplied. There will be new service—Waukegan Limited trains, making all Limited stops, half hourly in each direction; also Highwood Express trains, making all Express stops, half hourly each way. And there will be Local trains every half hour in both directions.

More Than Twelve Trains Every Hour

The foregoing schedule provides twelve trains every hour—six trains each way. In addition there will be 15 Milwaukee Limiteds daily—8 north-bound and 7 south-bound—over the Shore Line Route, all of them making the more important stops along the way.

The value of this frequent service is self-apparent. It should prove especially beneficial to Evanston, through attracting shoppers, from points further north, to this Shopping Center.

To Kenosha, Racine, Milwaukee

Fifteen Milwaukee trains each day, as previously explained, will operate via the Shore Line Route, making stops at Church Street and Central Street. These trains will provide direct, fast service from Evanston to Kenosha, Racine and Milwaukee. Persons wishing to travel via the Skokie Valley Route may do so conveniently by taking the Rapid Transit ("L") to Howard Street and boarding the North Shore Line there. The running time from Howard Street to downtown Milwaukee via the Skokie Valley will be 1 hour and 35 minutes.

Direct-Without-Change to Libertyville and Mundelein

With the opening of the Skokie Valley route, trains will operate direct-without-change from the Chicago Loop to Libertyville and Mundelein in the Lake County Countryside. Evanston people desiring to visit these places will find it convenient to board trains at Howard Street—a comfortable, pleasant 43-minute ride from there to Libertyville. Hourly service will be maintained in both directions.

Chicago North Shore and Milwaukee Railroad Co.
72 W. Adams St., Chicago

THE new Skokie Valley Route, costing nearly ten million dollars, is a model of modern railroad construction. From heavy rock ballast to steel arches carrying the power lines, it typifies the last word in transportation.

KEY
SKOKIE VALLEY ROUTE
SHORE LINE ROUTE

TO MILWAUKEE RACINE KENOSHA
WAUKEGAN
NORTH CHICAGO
LAKE MICHIGAN
MUNDELEIN LIBERTYVILLE
LAKE BLUFF
LAKE FOREST
SKOKIE MANOR
FORT SHERIDAN
SHERIDAN ELMS
HIGHWOOD
HIGHMOOR
HIGHLAND PARK
RAVINIA
BRIERGATE
GLENCOE
HUBBARD WOODS
WOODRIDGE
WINNETKA
NORTHBROOK
KENILWORTH
WAU-BUN
WILMETTE
GLENAYRE
HARMSWOODS
EVANSTON
DEMPSTER ST.
HOWARD ST.
WILSON AVE.
GRAND AVE.
CHICAGO LOOP
ADAMS
WABASH
ROOSEVELT ROAD
43RD ST.
63RD ST.

KENOSHA NORTH SHORE STATION
63rd St. and 27th Ave., Kenosha, WI

A North Shore Line *Electroliner* is alongside a Wisconsin Electric car being used for a "fan trip" at Kenosha, date unknown. (www.davesrailpix.com)

For more than 30 years, passengers from a variety of small Wisconsin communities could easily reach Chicago by making connections between buses and North Shore Line trains in Kenosha, WI. This type of traffic, together with Kenosha's sizable population, made the community one of the most important revenue generators on the North Shore Line.

The graceful brick station that opened in Kenosha in 1922 was considered one of the finest works of Arthur Gerber, the architect hired by North Shore Line's Samuel Insull to support a massive modernization effort. (Gerber's other work included Wells Street Terminal in downtown Chicago, which is also featured in this book, and the North Shore Line's Skokie station). Replacing an earlier and much smaller station, Kenosha's spacious one-story depot, situated on the Chicago–Milwaukee main line, had hourly departures in each direction and numerous connecting services, including local buses, streetcars, and intercity buses.

A particularly large number of passengers made transfers in Kenosha between the North Shore's trains and its buses operating to and from Lake Geneva, a popular service introduced the same year the station opened. Travelers making these bus-rail connections could do so using a single ticket. Others used North Shore buses operating over its shoreline route roughly parallel to its tracks.

By the early 1930s, the buses to Lake Geneva were operated by Cardinal Lines, an independent operator unaffiliated with the railway. In Bristol, Paddock Lake, Salem, Powers Lake, and other small Wisconsin towns, Cardinal's four daily Lake Geneva–Kenosha round trips were the only scheduled services of any kind available. Passengers could make transfers either at the North Shore Line station (the first stop in downtown Kenosha) or Union Bus Depot (the final stop). Union Bus Depot, housed in the former streetcar station at 55th Ave. and 8th St., was an important Northland Greyhound Lines stop.

Both stations were bustling places throughout World War II and into the early postwar era. In 1946, 37 trains and 8 intercity buses arrived and departed Kenosha's North Shore Line station each weekday. The railway's popular *Electroliner* streamlined trains covered the 52 miles to the downtown Chicago depot at Adams & Wabash in just 64 minutes.

By the mid-1950s, however, these bus and rail systems were in sharp decline. Bus service between Kenosha and Lake Geneva dropped from four round trips to a single trip in each direction in 1952, before being eliminated entirely in

A conductor assists a woman boarding an *Electroliner* at Kenosha, circa 1960. This image was taken from a Blackhawk Films production by A.C. Kalmbach, founder of Kalmbach Publishing, publisher of *Trains* magazine. (Author's collection)

the mid-1950s. The North Shore hemorrhaged red ink and finally suspended operations on January 21, 1963. In 1965, Union Bus Depot was replaced by a smaller station and torn down. Even the scant intercity bus service that remained could not be sustained indefinitely. After Kenosha lost its shoreline bus route linking it to Milwaukee in the early 2000s, service dropped to a lone daily Greyhound round trip between Chicago and Milwaukee, which would eventually be cut too. It was eliminated in January 2005, leaving downtown Kenosha without any long-distance service of any kind.

Kenosha, nonetheless, celebrates its transportation heritage with unusual zeal. Streetcar service has been established along the picturesque new shoreline loop. The community is home to a well-maintained commuter-rail station. As the endpoint of a heavily used Metra service, which runs over the former Chicago & North Western Railway route, its cultural and economic connections to the Windy City remain strong.

Gerber's classic North Shore Station was transformed into an Italian restaurant after the demise of its trains. That business failed, but the building was later restored to serve as a community center. Today, it is one of only three surviving North Shore Line stations, with the others being in Highland Park and Skokie. ∎

The Kenosha, station, shown here in 2012, appears frozen in time. Little has been changed from the 1962 photo, with the exception of the track bed and station parking lot, which are both gone. (Author's photo)

MICHIGAN CITY SOUTH SHORE LINE STATION
114 E. 11th St., Michigan City, IN

Several passengers prepare to board a westbound South Shore Line train during a 1981 stop at the 11th St. station in Michigan City, IN. This rare surviving "street-running" stretch of railroad has been the source of traffic delays for decades. (Mark Llanuza photo)

The South Shore Line's 11th St. Station in Michigan City was an important transfer point for passengers traveling to Chicago from smaller communities in central Indiana and southern Michigan. More than any other outlying point on the region's electric railway system, this station's service allowed for well-timed bus-to-train connections that were valuable to residents of many small towns outside the metropolitan region.

Service on this heavily used rail line through Michigan City began when the Chicago, Lake Shore, and South Bend Railway commenced operations in 1908. This interurban railway extended its routes in 1912 to Chicago following an agreement giving passengers through service over the Illinois Central Railroad between Kensington Junction, on the far south side of the city, and downtown stations. (South Shore Line cars were initially simply coupled to this mainline railroad's trains en route to downtown). The company, now having a 90-mile route from Chicago to South Bend, was reorganized as the Chicago South Shore & South Bend Railroad (South Shore Line) in 1925.

In 1928, the South Shore Line opened a two-story red-brick station designed by noted architect Arthur Gerber in Michigan City to serve the local market and provide greater comfort to passengers connecting between trains and buses linking this lakeside community and Benton Harbor. These buses, operated hourly by the South Shore Line, also made stops in New Buffalo, St. Joseph, Stevensville, and other smaller communities in southwest Michigan. Buses arrived and departed on the east side of the building while South Shore Line trains arrived and departed in front of the station in the middle of 11th St. Local transit operators also stopped at the station.

The transfer business at Michigan City gradually grew as other intercity bus lines began using the station, keeping it busy from dawn to dusk. In 1946, the Indiana Motor Coach Co. ran four daily buses to Indianapolis and another to Fort Wayne, IN, that originated at the station. Suburban Transit operated hourly service between the station and La Porte and Kingsford Heights, IN.

By the early 1960s, however, the transfer business was rapidly failing. South Shore Line's Benton Harbor bus service was sold to Indiana Motor Bus, which discontinued it in 1965. And the last intercity bus service of any kind serving the location—Indiana Motor Bus's daily round-trip from Indianapolis—disappeared in January 1981.

Today, the station is a registered historical landmark but is closed to passengers. South Shore Line riders are instead served by a modest glass-enclosed shelter adjacent to the old structure. ■

Thirty seven years after the previous photo was taken, an eastbound train of current-generation South Shore equipment passes the now-shuttered Michigan City station on September 3, 2012. By this time, passengers used a modest shelter just out of view, adjacent to the automobiles at right. The South Shore Line herald remains visible above the old station's front entrance. (Author's photo)

O'HARE BUS/SHUTTLE CENTER
Ground Floor Main Parking Garage, O'Hare International Airport

Coaches for Milwaukee and Champaign, IL,—destinations too close to Chicago to support low-cost air service—are loading under the O'Hare Bus/Shuttle Center's canopy at midday on June 10, 2012. Several hours later, starting at about 4 p.m., multiple buses will be needed at some destinations in order to accommodate peak-period loads. (Author's photo)

The O'Hare Bus/Shuttle Center has the most extensive network of intercity routes of any motor coach terminal at a major U.S. airport. The center has evolved from having only feeder routes from close-in points into an intercity hub with direct service to points in four states.

After its opening in 1955, O'Hare was most easily reached by automobile, taxicab, or CTA bus. Although the opening of the Northwest Expressway (today known as the Kennedy Expressway) in 1960 greatly improved the efficiency of taxicab and bus service from the city, ground-transportation service to the airport from longer distances was slow to materialize. The logjam opened in 1964, when Tri-State Coach lines launched limousine service to Gary and Hammond. By the end of 1967, Tri-State operated not only this service and other routes into the southern part of the region, but also departures for Milwaukee and Racine, WI. Buses departed O'Hare from the rotunda on the lower level of the airport roadway.

Eager to tap into this market, Greyhound began service from O'Hare in the late 1970s and even advertised its service to Milwaukee in the *Official Airline Guide* as an attractive alternative to flying. Although this service lasted only until 1980, the rising volume of bus and shuttle traffic into O'Hare grew to such an extent that it put a heavy burden on the airport's roadway system. The situation reached almost crisis proportions and continued to grow worse even after the CTA rapid-transit route (today's Blue Line) opened in 1984.

The following year, the city began charging most bus and shuttle operators a fee and required them to wait outside the terminal areas until their departure time was imminent. By the end of the decade, as many as 160 buses and vans operated hourly, leaving airport authorities scrambling for a solution. Concerns also grew that some bus passengers wandered through the airport simply to pass the time, adding to security costs. At one point, these passengers were asked to wait outside, even in bitterly cold weather, so that the terminals' interiors could be exclusively used for airline passengers.

A long-term solution to the congestion problem emerged as plans to build a new international terminal moved ahead. When this spacious new terminal opened in 1993, a ground-floor facility was transformed into a new bus and shuttle terminal. The area had once been part of the parking garage before being retrofitted as an overflow gate area for international flights.

Extensive renovations, including the construction of a bus-only roadway allowing for "run-through" operations, were made to permit the O'Hare Bus/Shuttle Center's opening in April 1996. Equipped with a large waiting room,

Passengers prepare to board a LEX Express bus at O'Hare.

ticket counter, and snack bar, the center began handling all bus and shuttle departures except those operated by car-rental companies and Continental Air Transport, which, by contract, were permitted to continue operating from the terminal. Bus and shuttle operators could still drop off passengers on the upper level of the airport roadway, providing curbside convenience for departing flights. Passengers arriving on domestic flights, however, had to proceed to the newly opened center.

Regional and intercity bus services from the center have grown to 80 scheduled weekday departures (and an equal number of arrivals) on six privately operated bus lines. This includes service by Act II Company to the Quad Cities; Express Coach to Davenport, IA; Coach USA to South Bend and Milwaukee; Lincolnland Express to Champaign; and Van Galder's service to Madison, WI. Greyhound, however, opted not to return and maintains a station at the Cumberland Blue Line "L" stop several miles away.

As a result of recent service expansions, the O'Hare Bus/Shuttle Center is acquiring some of the properties of a traditional intercity bus hub. Many arriving passengers transfer to the Blue Line trains destined for the Loop that depart one floor below. Others make connections between buses. In theory, a passenger today can travel all the way from Mishawaka, IN to Davenport, IA—a nearly 500-mile trip—with a connection at O'Hare. As noted in Appendix I, all services will eventually be relocated to the O'Hare intermodal transportation center, now under construction on Mannheim Rd. ■

Travelers pass the time at the O'Hare Bus/Shuttle Center in June 2012. Unlike the LCD departure monitors used elsewhere in the airport, departure information here is posted by the terminal's bus lines on printed signs of various shapes and sizes. The waiting room seats, however, resemble those found at a typical airport gate. (Author's photo)

TORNADO BUS COMPANY
2807 S. Kedzie Ave.

Passengers wait for a southbound coach in front of the storefront Tornado Bus station at 2807 S. Kedzie Ave. on June 2, 2012. This popular Latino-oriented carrier complements its basic transportation offerings with bilingual customer service. (John Dalide photo)

The Tornado Bus terminal on 2807 S. Kedzie Ave. is one of the northernmost points of a bus system, widely used by Spanish-speaking travelers on trips to and from Mexico, Texas, and the American South. Unlike other bus lines primarily serving the Latino population, which operate in motor coaches without obvious company identification, Tornado strives to build a strong company image with brightly painted buses and stations.

The Chicago station consists of an air-conditioned waiting room with a ticket counter. Buses depart from the curb in front of the depot, and vans painted in the company's color scheme shuttle passengers between the station and residential neighborhoods—a service included as part of the ticket price.

Service to this station began in 2006 with a single daily trip to and from Dallas, one of the company's primary hubs. Since then, a large network of routes through the Deep South has been developed. Generally two, but sometimes up to four, daily buses from Dallas stop at this South Side station before continuing to Waukegan and Milwaukee. By having two drivers onboard, buses are able to operate on relatively fast schedules. The trip to Dallas takes 16 hours, compared to about 20 on Greyhound, while the two-segment trip to Mexico City requires about 30.

A competitor to Tornado, the El Expresso Bus Co. (see Appendix II) maintains a smaller station at 3501 S. California Ave., about a mile and a half away. This rival was purchased by Tornado in 2012 but maintains a separate identity. Both are part of an expanding network of bus lines well known in the Latino community, but unfamiliar to many other travelers. ■

Catering to its target clientele, the station is equipped with a ticket counter, a large map of Mexico, and a television tuned to a Spanish-language channel. (Author's photo)

Passengers tow their luggage outside a Tornado Bus coach that will make the 16-hour trip to Dallas—the carrier's primary route from Chicago. At least three buses leave this station for the Texas hub on the busiest days, with most fares paid in cash. (John Dalide photo)

WOODFIELD MALL BUS STOP
Parking Lots along Perimeter Drive, Schaumburg, IL

A LEX Express bus departs for Urbana–Champaign, IL, from the carrier's unmarked location at the southwest side of the Woodfield Mall on July 29, 2012. By using hand-me-down coaches, the campus-oriented carrier was able to hold fares at less than $30 on this route. After being charged with safety violations, though, this carrier shut down several months later. Several other operators remain. (Author's photo)

Woodfield Mall has been an important terminus for intercity buses serving major university campuses for more than 25 years. On peak-demand days, upwards of 30 buses arrive and depart from this megamall's parking lots, carrying passengers to campuses throughout the Midwest.

The genesis of these services can be traced to the mid-1970s, when the Village of Schaumburg was experiencing spectacular growth. Situated near the interchange of the Northwest Tollway (I-90) and I-290, the community had expanded from just 3,000 residents in 1960 to more than 32,000 in 1970, making it an attractive market for intercity bus companies. The need to serve Schaumburg was imperative after Woodfield Mall opened in 1971. Within two years, this massive retail complex boasted 159 stores and was the largest mall in the United States.

Within a few years, both Greyhound and Trailways were eyeing potential station sites within or near the I-90/I-290 interchange. Greyhound moved in first by establishing a stop at the Park & Shop Shopping Center in Elk Grove Village, approximately four miles east of the interchange, in September 1978. By virtue of its location along the carrier's Chicago–Madison, WI, route, this stop, situated near Elk Grove Village's massive business park, became an important source of freight and passenger revenues.

Finding the revenue from transporting college students to be lucrative, Greyhound launched weekend bus service from Elk Grove Village to the University of Illinois at Urbana–Champaign and Illinois State University (ISU) in Bloomington. These buses, departing for campuses on Sunday evenings and returning on Fridays, and making intermediate stops at Oak Brook Mall and other suburban locations, proved quite popular.

A headline in the *Daily Herald* on October 20, 1985, summarized the David-and-Goliath battle for market share on the Woodfield–Champaign market.

Fare wars
Bus service run by student butts heads with Greyhound

Aware of the success of Greyhound's services, Trailways established a stop at a travel agency in Elk Grove Village in 1983.[50] A third operator, Suburban

Express, entered the fray by inaugurating service to Urbana–Champaign later that year. This company, founded by a former Greyhound rider who considered the national carrier's fares to be excessive, set out to prove his point by undercutting it. Operating from a corner of the Woodfield Mall parking lot to Urbana–Champaign, Suburban Express used hand-me-down buses on peak days and lower-capacity vans on slower-demand days.

Greyhound matched Suburban's prices and established its own stop for campus buses from Woodfield. A "fare war" erupted, much to the delight of the college students, unaccustomed to such attractive pricing.[51] More and more students, however, shifted their loyalties to Suburban Express, in part due to its underdog status. Greyhound's losses were so great that it closed its Elk Grove Village station and abandoned service there in mid-1990, before dropping the last of its Woodfield service in late 1991.

Schaumburg's role as a hub for campus bus service, however, continued to grow. In 1999, Lincoln Land Express, called "LEX Express," began a competing campus oriented bus service to Urbana–Champaign on the northwest end of the parking lot. LEX also inaugurated service to ISU and more distant colleges by way of a connecting hub in Urbana–Champaign. Within a few years, other carriers also began service, giving students Sunday departures during the school year to out-of-state campuses, including Indiana University, Purdue University, and the University of Notre Dame. Students who previously could only reach their campuses by making time-consuming transfers in downtown Chicago flocked to these new services.

By early 2012, the Woodfield operation had grown to encompass more than 30 daily departures on peak days and was attracting increasing numbers of passengers who were not registered students. Later that year, however, LEX was charged with a series of safety violations and opted to shut down, increasing Suburban Express's dominance and reducing the number of daily departures by about a third. ■

A Suburban Express coach, identifiable only by the yellow sign next to the door, boards at Woodfield Mall in 2012. Founded in 1983 by a student disgruntled with the existing bus service, this carrier effectively drove Greyhound out of the market from Champaign to the western suburbs. This bus stop is devoid of signage and is a considerable distance from the nearest mall entrance. (Author's photo)

III. AIR TRANSPORTATION
Terminal Town

AIRPORTS
Terminal Town

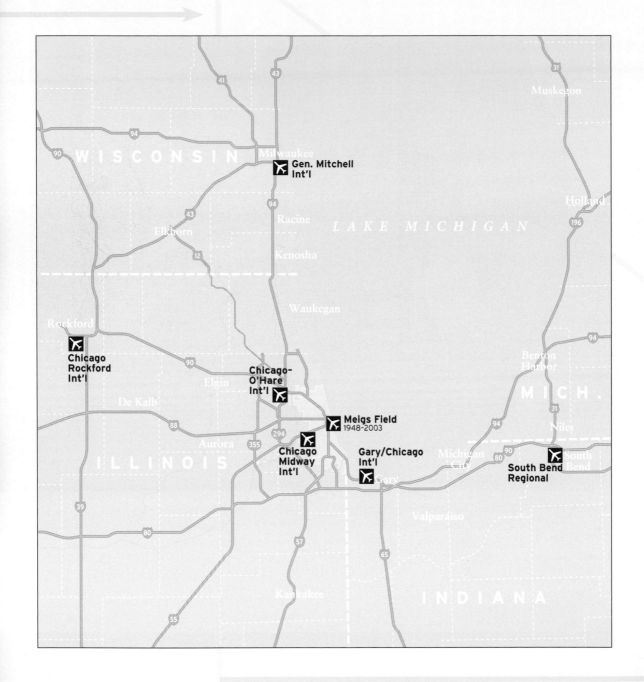

An Allegiant Airlines Super-80 is pushed by a service vehicle at Gary/Chicago Airport on August 10, 2013. The plane's departure would mark the end of scheduled services from the airport. (Author's photo)

Perhaps no single word in transportation representing a particular geographic place is as recognizable to Americans today as "O'Hare." In much of the United States, the same could have been said of "Midway" sixty years ago. Simply put, in the days before Caravelles, DC-8s, and 707s, Midway (formerly Municipal) Airport was among the most significant square miles of land devoted to commercial aviation in the world. As jets replaced propeller-driven planes, the locus of activity shifted to the newly opened O'Hare International Airport, which for many years was in a category all to its own.

A few statistics underscore Chicago's long-standing status as the country's premier air-travel hub. The number of destinations served nonstop by scheduled passenger airlines from airports

(Opposite) This map shows the seven airports serving the Chicago region that had intercity passenger service for sustained periods between 1939 and the present. Many residents in outlying suburbs today travel outside the metropolitan area to use airports in Milwaukee, Rockford, and South Bend as an alternative to Midway and O'Hare. Only Milwaukee's General Mitchell Airport has been continuously served by scheduled (non-charter) airlines since 1939. Meigs Field was closed by the City of Chicago in 2003.

within the City of Chicago has had no equal in American aviation for more than 70 years. Municipal (Midway) Airport was the world's busiest between 1948 and 1961. O'Hare Airport held this distinction from 1962 until 1998. Today, airports in *metropolitan* New York handle more passengers than those in the Chicago region, and Atlanta Hartsfield Airport sees more air travelers than O'Hare, but more passengers board flights at airports in Chicago—at Midway and O'Hare combined—than from any other city in the United States.[52]

Airports in the Chicago metropolitan area that lie beyond the city limits, have fared poorly in the development of scheduled service. None has consistently sustained service to points outside the metropolitan region. The construction of a third major airport to ease congestion at Midway and O'Hare has been "on hold" for years, partly because of political issues.[54]

The enormous costs and complexity of modern airport development are a big part of the story. Years ago, suburban governments lacked the financial and technical sophistication to embark on large-scale airport projects that could effectively lure significant airline traffic away from the city. The absence of viable alternatives to Midway and O'Hare has for many decades saddled the region with an imbalance in the supply and demand for airport capacity. Despite recent improvements made to O'Hare, Chicago is still seen by some as one of America's worst aviation bottlenecks.

The problem is not that suburban airports have failed to capture the interest of airlines. Gary and Valparaiso, IN, as well as DuPage County Airport, Sky Harbor in Northbrook, IL, and several others have had scheduled service to points outside the Chicago region since World War II. Rather, the problem has been that these airports have never been able to achieve the critical mass necessary to compete with their giant counterparts in the city.

Even so, O'Hare's creation was the crowning achievement in the development of the city's intercity transportation system during the early post-World War II era. Designed to relieve congestion at hemmed-in Municipal Airport (renamed Midway Airport in 1949) and handle larger and longer-range aircraft that needed longer runways than were available at this older facility, the airport was built on an unincorporated parcel of land situated between Bensenville, Elk Grove Village, and Rosemont, IL, making it suburban in orientation. Yet this property was brought under city control through a complex—and highly political—process.

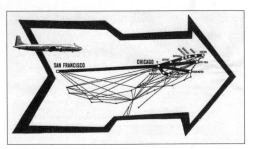

An American Airlines Chicago–San Francisco advertisement from 1956 points the way to understanding Midway Airport's enormous role in the development of east-west travel across the United States. (Author's collection)

Construction of O'Hare took years, and it was not until 1955 that it saw its first commercial flights. This massive multi-runway facility was an enormous success, causing Midway to gradually surrender its air-travel role. As Midway declined, tiny lakeside Meigs Field, built on a landfill near the south end of Grant Park and catering to downtown business travelers, began to absorb more flights.

By late 1962, Midway was completely devoid of

(Opposite) A poster advertisement from the 1960s links "Chicago's hometown airline" to the splendor of Marina City, at the time an exciting new Chicago landmark. Years later, United moved its company headquarters to 77 W. Wacker Dr., directly across the river from the corn-cob-shaped towers. (Author's collection)

scheduled flights except for those of a helicopter operator and charter companies. The airport nonetheless found a champion in Mayor Richard J. Daley, who, convinced that much more airport capacity would be needed, pushed hard to save the facility, which was largely abandoned. The mayor, buoyed by the success of Meigs Field and following his penchant for thinking big, similarly explored the possibility of an enormous airport on a man-made island in Lake Michigan as well as one on the far South Side of the city at Lake Calumet.

Limited scheduled-flight activity at Midway resumed in 1964, but the debate about the need for more airport capacity continued to swirl—and it rose sharply in intensity after the airline industry's deregulation in 1978. Congestion at O'Hare escalated, and a titanic struggle for dominance between American Airlines and United Airlines suggested that the construction of new runways, whether at O'Hare or somewhere else, was urgently needed to avoid "gridlock." Flight limitations were introduced, but the deregulation of the airline industry led to such traffic growth that O'Hare's shortcomings became a matter of national concern. A resurgent Midway also shouldered an increasingly heavy burden, but growth was so robust in the late 1980s that the late mayor's son, Richard M. Daley, vigorously supported a coalition championing an old idea: building a new airport at Lake Calumet.

The effort to build the Lake Calumet airport ground to a halt in 1992, having been abruptly dropped by Daley because of concerns over who ultimately would control the facility, environmental issues, and the need to expropriate large numbers of homes and businesses. Fortunately, Midway absorbed more of the demand than was previously envisioned, due in part to the popularity of service offered by Southwest Airlines, and gave the region time to assess its options.

The failure of the Lake Calumet scheme, however, did not sit well with the suburbs and set in motion a contentious debate about whether to expand O'Hare or build a new airport in south-suburban Peotone, about 40 miles south of the Loop. The City of Chicago, having long been able to dominate decisions about airport planning, was adamant that funds from Midway or O'Hare could not be used to support the Peotone project. In 2001, the city embarked on the massive O'Hare Modernization Program (OMP) despite mounting opposition from nearby suburbs. A coalition of suburban governments, meanwhile, championed the viability of the "South Suburban Airport" at Peotone.

Despite the weakness of air travel after the 9/11 terrorism acts, Mayor Daley never wavered in his support for the OMP, and his team silenced many skeptics by finishing the initial phase of the project within budget. Nevertheless, new competitors to Chicago's airports emerged as airline trips became longer and more-oriented toward pleasure travel. Airports in Rockford, IL, and South Bend, IN, enjoyed some success attracting travelers from the Chicago area, while Milwaukee's Mitchell Airport began attracting passengers in large numbers.

For promoters of Chicago's existing airports, however, these satellite facilities still posed less of a threat than the proposed South Suburban Airport in Peotone. But the Peotone proposal became embroiled in controversy: not only was there opposition from people living near the proposed site, but also questions over political control divided the plan's supporters. A massive state-sponsored land-acquisition program has not yet been enough to move the effort beyond the proposal stage.

The airport in Gary, renamed Gary/Chicago International Airport, also has the potential to help alleviate the region's air-travel problems. The prospects for Gary's airport further improved when the State of Indiana strengthened its commitment to lengthen the runway in 2010. Few believe, however, that an expanded Gary Airport, which has seen many airlines come and go, will play a major role in scheduled passenger service anytime soon.

As is detailed in the following pages, the evolution of the airport system serving residents of the Chicago region has taken many surprising turns since 1939. All of the featured airports (presented in the approximate order in which they became important to the Chicago region) have had scheduled service to points outside the Chicago region for periods of five years or more. Several others, mostly limited to providing air-taxi service to Midway or O'Hare, are discussed in later chapters. ■

One of the two baggage-claim carousels at Chicago Rockford International Airport is lined with passengers who have just arrived from Las Vegas on Allegiant Air. In contrast to years ago, when Rockford had frequent service on mostly short-hop routes, today the city is served by 160-seat jets flying to distant leisure destinations but with considerably less frequency, thereby creating a boom-and-bust cycle in the terminal. (Author's photo)

CHICAGO MIDWAY INTERNATIONAL AIRPORT
5700 S. Cicero Ave.

Midway Airport's classic 1947–48 terminal building and control tower, while appearing relatively quiet in this May 1997 photograph, was being pushed to the limit by the rapid expansion of Southwest and ATA Airlines. The area inside the circular window pattern below the control tower once housed the posh Cloud Room restaurant, a popular haunt among celebrities and wealthy travelers. (Lawrence Okrent photo)

CHICAGO MIDWAY INTERNATIONAL AIRPORT
5700 S. Cicero Ave.

The Midway Airport "South Terminal," built in 1931, seems frozen in time, circa 1976, nearly 30 years after flight activity shifted to the much larger terminal a half mile farther north along Cicero Ave. While sporting an elegant exterior and an interior with distinctive Art-Deco themes, this two-story landmark had no place in the airport's long-range plans and was demolished a short time later. (Pat Bukiri photo)

(Opposite) With the boarding stairs rolled away and propellers spinning, an Eastern Air Lines Prop-Jet Electra appears ready to depart at Midway on September 7, 1960. A boxy concourse extension can be seen in the foreground—a testimonial to the ad hoc nature of expansion efforts while traffic surged. (Jon Proctor photo)

Like a chameleon, Chicago Municipal Airport (renamed for the World War II Battle of Midway) has had to adapt to meet the challenges of a changing environment in order to survive. This airport was the world's busiest airport throughout much of the 1930s and again from 1948 to 1960.[56] Midway's utility and luster faded as airlines moved their flights to the newer and much-larger O'Hare International Airport in the early 1960s, and it was briefly at risk of abandonment. The airport eventually rebounded by offering short-haul flights to business travelers before becoming one of the country's premier discount-airline hubs.

Initially built on a rectangular 320-acre tract of farmland, this airport on the city's Southwest Side saw its first flights in 1922, and was dedicated as Chicago Municipal Airport in 1927.[57] As travelers across the country embraced the convenience of scheduled flights, service at the new Chicago airport grew prodigiously. In late 1929, Municipal had 26 daily flights offering nonstop service to 10 cities, with the farthest such flight reaching Kansas City. Multiple-stop service was available to many more destinations.[58]

A new terminal by architect Paul Gerhardt, equipped with a control tower, a small waiting room, and notable Art-Deco features, was opened in 1931. The footprint of the airport was gradually expanded to one square mile. By 1940, the small terminal was being stretched to the limit, handling 122 daily flights and offering nonstop service to 28 destinations, the furthest of which was Denver.[59] Travelers destined to Florida and California, however, still had to make three or more stops, and Seattle flights generally required four.

Flight activity boomed again after the country entered World War II. An economic recovery coupled with wartime demands and the introduction of larger and longer-range aircraft stimulated traffic to such an extent that the airport's parking lot spilled over to the east side of Cicero Ave. and the city searched for ways to expand the terminal.

With expansion plans delayed by other wartime priorities, the 1947 opening of a new terminal, several times larger than its predecessor and located farther north along Cicero Ave., generated considerable fanfare. This classically designed terminal, also by Gerhardt, partially wrapped around a new parking lot and had its control tower at the center. Two new concourses were added in 1952, but even this proved inadequate to deal with all the traffic growth. By 1955, the number of flights serving Midway rose to more than 400, and passengers could fly nonstop to places as far away as Los Angeles, San Francisco, and Seattle. The total number of cities served by nonstop flights then stood at 52, not including destinations served by air-taxi operators that shuttled passengers between Midway and the city's suburbs.

Advances in aviation technology became so rapid that the entire airport was threatened with obsolescence. The introduction of the first generation of jet airliners, such as the Boeing 707 and Douglas DC-8, required longer runways than were available at Midway. As early as 1956, airlines had begun relocating

An illuminated "Welcome Back" sign, installed on behalf of Chicago Mayor Richard J. Daley, greets pilots and passengers arriving on Midway's main northwest-southeast runway after United Airlines resumed service in 1964. (Author's collection)

CHICAGO MIDWAY INTERNATIONAL AIRPORT
5700 S. Cicero Ave.

Midway Airport remains a busy place on September 7, 1960. An Ozark DC-3 is facing the camera while an American Airlines Electra is on the opposite side of the concourse. Several TWA Super Constellations, capable of flying (like the Electras) nonstop to the West Coast, are visible in the distance. Within two years, these three carriers had pulled up their stakes and moved all flights to O'Hare. (Jon Proctor photo)

long-haul piston-engined flights to newly opened O'Hare Airport on the city's northwest side. A helicopter service established that year began shuttling passengers between the two airports.

As late as 1960, however, Midway still was a major player on the national aviation scene. Well over 400 daily flights still whisked passengers nonstop from Midway to 65 cities—a greater number than even before O'Hare's opening in 1955. But Midway's market position was eroding, and by late 1961 airlines had abandoned it in droves. The 1962 opening of the Northwest Expressway (today known as the Kennedy Expressway) further diminished Midway's role by making O'Hare much easier to reach. In late 1962, the once-unthinkable outcome—the loss of all scheduled flights, except for the helicopter service to the city's other airports and those of supplemental (e.g., charter) air carriers operating to points, at most, a few times a week—overtook Midway.

Chicago Mayor Richard J. Daley insisted that the city fight to "save Midway" and pushed airlines to return.[60] United Airlines resumed scheduled service in July 1964, in part due to a city requirement that airlines flying to O'Hare also serve Midway and improved access resulting from the opening of the Southwest Expressway (today's Stevenson). A modest recovery took root as traffic spilled over from an increasingly congested O'Hare. By 1970, Midway had 76 daily flights operating nonstop to 32 cities—a far cry from the number available a decade before, but nonetheless a significant improvement from only a few years earlier.

By the mid-1970s Midway's fortunes had again turned for the worse. The number of flights ebbed, in part due to rising fuel costs, and signs of neglect were everywhere. Clumps of grass sprouted from cracks in the runways. The debate about whether to permanently close the airport—which only had service to St. Louis and a handful of other cities—and redevelop the property began anew. As pessimism about the airport mounted, a ray of hope emerged following the deregulation of the airline industry in the form of new Midway Airlines flights to Cleveland, Detroit, and Kansas City. These were the only scheduled passenger flights serving Midway when they began in 1979, but the discount carrier gradually expanded. After Southwest Airlines launched service in 1985, a full-blown renaissance began, a process enhanced by the opening of the CTA Orange Line rapid-transit link to Midway in 1993.

By 2000, nonstop flights were available from Midway to 38 cities and the airport had once again become a star performer. Midway fared better than

Even as passengers increasingly favored air travel over passenger trains during the mid-20th century, support functions found in airports—such as waiting areas, baggage handling, and restaurants—had much in common with those provided in the nation's railroad terminals. Here, passengers dressed for business wait in the American Airlines concourse at Municipal Airport, circa 1955. (Chicago History Museum; ICHi-26487)

A Southwest 737-700, sporting the company's familiar colors, taxis against the backdrop of the John Hancock Center and Sears Tower on May 25, 2008. Several more Southwest 737s in the distance attest to the carrier's growing dominance at Midway. (Brian Futterman photo)

most other U.S. airports after the terrorist acts of September 2001, partly due to aggressive expansion by discounters Southwest and the airport's second-largest carrier, ATA Airlines. As each of these airlines expanded its hub, a complete overhaul of Midway's passenger facilities became essential, culminating in the 2004 opening of a new terminal, parking garage, and passenger concourses. The new terminal building built on the east side of Cicero Ave., designed by HNTB, made its predecessor seem tiny in comparison. The glass-enclosed terminal was

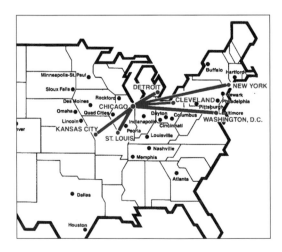

A Midway Airlines route map from its October 26, 1980, timetable shows the carrier's three original routes and ambitious expansion plans. (Author's collection)

Fight Over Midway's Restoration Looming

A *Chicago Tribune* headline on August 12, 1964, informed readers of Midway Airport's uncertain fate. (Author's collection)

linked by a spacious pedestrian bridge to the passenger concourse built on the site of the old main parking lot. Not a trace of the esteemed 1947 passenger facility remained, making this bygone landmark one the most recent examples of a large and historical Chicago terminal to be demolished.

Today, Midway boasts more than 500 daily flights and nonstop service to 85 destinations, including international routes to Canada and Mexico. Passenger traffic is at an all-time high. Southwest and its subsidiary AirTran account for 91% of daily enplanements at this resurgent airport.[61]

Yet, in many ways Midway is an anachronism. Its footprint, still measuring barely more than one square mile, cannot be feasibly expanded. Its runway configuration, while adequate, is now sorely outdated. Aircraft fly uncomfortably close to the city's residential "bungalow belt." Aware of its remarkable legacy, members of an informal organization, the Midway Historians, meet regularly to study the many phases of its past.

Midway's robust performance in recent years and its secondary role to O'Hare are qualities that make it a possible candidate for long-term lease to a private concessionaire. Like a chameleon, Midway seems destined to continue its adaptation to the changing times.[62] ■

This view from the upper floor of Midway's terminal building shows how a vertical design helped minimize the problems posed by a hemmed-in location. Airline ticket counters and check-in areas are on the top floor; the middle floor provides access to the parking garage and CTA trains, passenger concourse and security lanes. Baggage carousels are located on the ground floor. (Xhoana Ahmeti)

This image, one of the many thousands from the extraordinary aerial photography archive of veteran Chicago planning and zoning consultant and aerial photographer Lawrence Okrent, shows Midway's tarmac and passenger concourses dominated by planes painted in Southwest Airline's signature colors—orange, purple, and red. This September 15, 2007 photo also shows the airport's intimate relationship to the city's Bungalow Belt. The CTA Orange Line station is at right. Note the connecting bridge between the main terminal building on the west side of Cicero Avenue and the concourse areas—a hallmark of the airport's design. (Lawrence Okrent photo)

O'HARE INTERNATIONAL AIRPORT
10000 West O'Hare Ave.

An American Airlines 737-800 is taxiing at O'Hare Airport on April 21, 2012, five months after the company declared bankruptcy. As part of its restructuring, American opted to keep its Chicago operation at full strength, in part due to its battle for market share with United. The flat-topped 1962 terminal building can be seen in the distance. (Boeing 737-823 American Airlines N925AN, by Magic Aviation, is licensed by CC BY-SA 2.0)

More than any other transportation hub, O'Hare International Airport exemplifies Chicago's preeminence on the American transportation scene. Between 1961 and 1998, O'Hare owned the title of the world's busiest airport. Among U.S. airports today, it ranks behind only Atlanta with respect to both passenger traffic and operations.

Established primarily to complement the much smaller Midway Airport on the city's southwest side, O'Hare was built partially on decommissioned land formerly occupied by Orchard Place Field, an airport built by Douglas Aircraft in 1943. This property was brought under city control through a complex—and highly political—process that required annexing a thin sliver of land that later became the Northwest Expressway (today's Kennedy Expressway).

O'Hare opened in several phases after construction started in the late 1940s. Scheduled service began with an American Airlines flight from Detroit on October 30, 1955.[63] Twenty flights used the airport on its opening day, and service gradually expanded as airlines moved flights to it from Midway, where runways were too short for the emerging first generation of jet airliners. By late 1956, O'Hare had nonstop service to Los Angeles and San Francisco, and Chicago Helicopter Airways operated a popular shuttle service to Midway and downtown Chicago. By late 1957, the number of nonstop destinations had risen to 18, setting the stage for the opening of a temporary international arrivals and customs facility in 1958. Nonstop flights to Europe began in 1958 with Trans-World Airlines' Starliner Constellation service to London and Paris. Jet service across the Atlantic began in 1960 with German airline, Lufthansa's, launch of Boeing 707 service to Frankfurt.[64]

Despite this start, development of O'Hare was far from complete. Midway Airport still had service to almost twice as many domestic destinations as O'Hare in 1960 (65 vs. 34).[65] Most passengers flying on shorter-distance routes, where jets were less common, still used Midway. After the Northwest Expressway and I-90 extension (linking the airport to downtown Chicago) opened in late 1960, however, the stage was set for O'Hare's eventual dominance. A new steel-and-glass terminal complex designed in the then-pervasive International Style by architecture firm C.F. Murphy was opened to the public in 1962, while a new two-level roadway—a prototype for big-city airports—permitted convenient passenger drop-off and pickup. A "rotunda" between the two main terminals offered dining options, ground-transportation services, and splendid views of arriving and departing aircraft and ramp activity.

All flights serving Midway had been relocated to O'Hare by the end of 1962. The number of carriers serving O'Hare gradually expanded, but United Airlines, followed by American Airlines, grew particularly fast. Congestion grew so rapidly that the federal government imposed slot controls in 1969 that

Traffic is moving freely on the upper level of Terminal 1's roadway at O'Hare on Sunday evening, June 30, 2013. The roadway, viewed here from the rooftop of O'Hare's main parking deck, has been widened to six lanes, completely obscuring the arrivals level below. (Author's photo)

O'HARE INTERNATIONAL AIRPORT
10000 West O'Hare Ave.

A Capital Airlines Vickers Viscount and nearly brand-new Continental 707, as seen from the O'Hare's observation deck in September 1959, are ready for boarding. Capital, a major player in the Chicago–East Coast market, was merged into United Airlines in 1961, setting the stage for the development of United's large local hub. (Jon Proctor photo)

limited the maximum permissible number of daily takeoffs and landings. Even with these restrictions, in 1970 O'Hare was a transportation behemoth, with nonstop service to a staggering 130 North American destinations, plus extensive transoceanic service. The opening of a new runway in 1971—the last one to be built at O'Hare for more than a quarter-century—and construction of a massive parking lot (purported to be the largest in the world at the time) helped fuel another round of growth.

United Airlines moved quickly to establish a much larger hub at O'Hare after deregulation of the airline industry in 1978 gave it additional freedoms, and American Airlines soon tried to match its competitor's every move. Many short-haul routes (to points within 100 miles of Chicago) were abandoned so

The belly of a United Airlines 767-300 is being loaded with cargo on June 24, 2011. This Europe-bound jetliner, like most of the airline's transatlantic departures, has been assigned to the "C" concourse, putting it a considerable distance from the main ticketing area, as is evident by the floodlights on the horizon. (United Airlines B763 N651UA, by Lasse Fuss, is licensed by CC BY-SA 3.0)

that their slots could be freed up for more profitable flights to points beyond the Midwest, resulting in new nonstop service to more-distant markets that could previously be reached only on one-stop trips.[66] Efforts to expand O'Hare's terminals and "groundside" transportation services kicked into high gear as the airport's global importance increased. Chicago Transit Authority rapid-transit service linking the airport with downtown began in 1985, the all-new Terminal 1 opened to serve United in 1987, and the massive Terminal 5 international complex opened in 1991. O'Hare's increasing dominance in air service to many midsize Midwestern cities, together with its extreme vulnerability to bad weather, made it, inevitably, the target of criticism.

The era of uninterrupted growth in passenger boardings did not last indefinitely, however, and this traffic peaked in 1998, when O'Hare relinquished its status as America's busiest airport to Atlanta. After the terrorist acts of 2001 sharply reduced demand for air travel, both hub operators scaled back and gradually substituted regional airliners for larger jet aircraft. Nonstop service to some smaller cities vanished. International service, meanwhile, remained a bright spot due to the strength of American and United's global alliances.

O'Hare was still severely congested despite the drop in domestic service, setting into motion a contentious debate about the best way to expand the airport. Delays were particularly bad during times of inclement weather, giving O'Hare the reputation of being the nation's worst aviation bottleneck.

A United Airlines Viscount appears to be receiving maintenance on January 31, 1963. A mechanic has just climbed through the emergency exit onto the wing. Two Eastern Air Lines Boeing 720s and a DC-7, together with one of Continental's unusual tubular loading bridges (upper left), are at the opposite concourse. (Jon Proctor photo)

10000 West O'Hare Ave.

Passengers stroll through United Airlines' Concourse C, one of two parallel concourses designed by Helmut Jahn at O'Hare and opened in 1987. Its appearance has changed little since this 1991 photo was taken, although large stacks of newspapers are only a memory. (Concourse C from 1991, by Phillip C, is licensed by CC BY 2.0)

O'HARE INTERNATIONAL AIRPORT
10000 West O'Hare Ave.

A passenger with a child gazes at a group of United Airlines regional jets at Terminal 2 in June 2013, not long after the 50th anniversary of the dedication of the terminal. This glass overlook is virtually unchanged since 1962, although it is now behind security and inaccessible to visitors without tickets. (Author's collection)

Congress warned that unless local officials worked to solve the problem, the federal government would, in effect, take action on their behalf.

A new runway and control tower opened in 2009, the first phase of a massive expansion project known as the O'Hare Modernization Program. Mayor Richard M. Daley, the airport's most vigorous champion, turned his attention to the program's "completion phase," which necessitated demolition of hundreds of homes and many businesses, with a particularly large number in suburban Bensenville at the facility's southwest edge. Work continued after Daley left office, and in 2012, Irving Park Rd. was relocated to permit runway construction on the south side of the airport. Another major milestone was reached in 2013 with the opening of a new east–west runway on this same side of the airport.

O'Hare today has nonstop service to more North American cities than any other airport in the country and stands alone among the world's airports in serving as a full-scale connecting hub for two global giants: American Airlines and United Airlines. As noted in Appendix I, several improvements envisioned for O'Hare could one day change the way residents reach this massive facility. ■

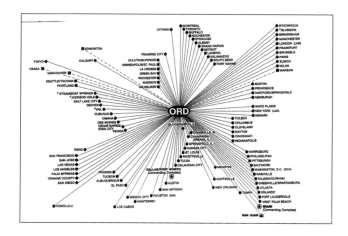

This stylized map appearing in an American Airlines 1982 timetable shows the company's O'Hare hub near its maximum extent. A careful review of the map shows the relative strength of American's O'Hare services to European and North American destinations compared to points in Asia and Latin America. (Author's collection)

Passengers ascend and descend to the main concourse level of Terminal 3 at O'Hare. The check-in lines at left are part of American Airlines expanding international network that sees a surge flight activity in early evening as outbound flights depart for Europe. (Xhoana Ahmeti).

O'HARE INTERNATIONAL AIRPORT
10000 West O'Hare Ave.

This photo, previously appearing in Lawrence Okrent's *Chicago From the Sky: A Region Transformed,* shows the vast O'Hare terminal complex on April 16, 2006. Virtually the entire complex, except for the International Terminal in the upper right, appears to be experiencing heavy traffic. The architecture of the airport's original black-roofed terminals and concourses contrasts sharply with United Airlines' glistening Terminal 1 (lower left). (Lawrence Okrent photo)

O'HARE INTERNATIONAL AIRPORT
10000 West O'Hare Ave.

Travelers using the moving sidewalk linking the twin concourses at Terminal 1 are treated to a lively display of neon light on May 26, 2009. The walkway's colorful illumination has earned accolades from architects for serving to relieve some of the stress associated with flight transfers at the massive airport. (United Airlines corridor, Chicago OHare Airport (6196116901), by InSapphoWeTrust, is licensed by CC BY-SA 2.0)

This United Airlines route map shows the extent of its O'Hare hub in 2013. The dense cluster of routes emanating northeast from Chicago is the carrier's expanding transatlantic operation; routes stretching toward northwestern Canada link Chicago to the Pacific Rim. The relative paucity of Latin American routes (São Paulo, Brazil, was the only South American city reached nonstop at the time) remains one of the few weaknesses of O'Hare's vast route network. (Author's collection)

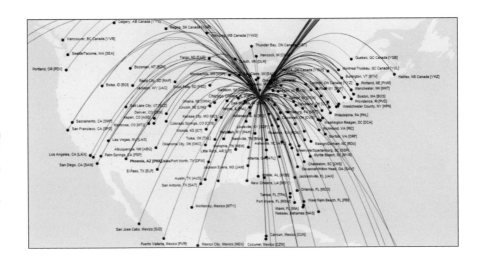

A TWA jumbo jet is on an O'Hare taxiway, circa 1972. After launching 747 service in 1970, the carrier strengthened its O'Hare service by flying these planes between Chicago and both Las Vegas and London. In 1991, TWA sold the latter route to American, which continues to fly nonstop to Heathrow Airport, albeit with smaller twin-engine planes. (Jon Proctor photo)

An American Airlines agent in Terminal 1 at O'Hare points the way for a customer who apparently has a long walk ahead of him. This traveler is likely going to the L concourse on the far east side of the terminal. Terminal 3's separate concourses for G, H, J, and L gates elicit confusion among some flyers. (Xhoana Ahmeti photo)

GARY/CHICAGO INTERNATIONAL AIRPORT
6001 Airport Rd., Gary, IN

An Allegiant Air MD-80 at Gary/Chicago is being serviced for its departure to Orlando Sanford International Airport on July 6, 2013. Barely a month later, on August 10, 2013, Allegiant suspended its Gary service. (Author's photo)

Gary/Chicago International Airport is the least-known member of an exclusive club in the region: it is one of just three airports (with the others being Midway and O'Hare) that has had scheduled passenger service provided by jet aircraft in the post-World War II era. (The other airports with jet service often used by area travelers, such as Rockford, are outside the region's official boundaries). Formerly named Gary Municipal Airport, this facility has seen many carriers come and go over the past 60 years, demonstrating a tenuous commercial viability. Recent developments, however, suggest that it may soon play an expanded role.

Since scheduled passenger service began at this airport in 1953, it has been provided in two distinct phases. The first phase emphasized short-haul flights, mostly to Midwestern points. The second phase featured more long-haul, albeit relatively infrequent, flights to markets in Florida, the Carolinas, and the American Southwest. These services were primarily attractive to tourists.

The original Gary Airport was a mere grass strip at 61st and Broadway St. in what is now Merrillville and replaced by the present airport, which is situated farther west on land that had been purchased in 1939 and cleared for a synthetic rubber plant, which was never built. After a paved runway was laid down in 1949, and other improvements, including a beacon, field lighting, and runway extension, were made, Lake Central Airlines in February 1953 introduced Gary as a stop on flights between Midway Airport and Indianapolis as well as Lima, Ohio, and Pittsburgh.[67] Airport officials publicized the fact that Lake Central's DC-3 airliners reached Midway in just 15 minutes, allowing for easy connections to transcontinental flights. A second carrier, Chicago Airways, made Gary a stop on its Midway–South Bend route in 1955.

These services, however, performed poorly. The Chicago Airways flights had been eliminated by 1957, and Lake Central service reportedly averaged fewer than five passengers daily in Gary, prompting Lake Central to suspend service to Gary in January 1958, and permanently eliminate the stop the following November. Chicago Helicopter Airways partially filled the void by launching three daily departures from Gary to Midway and O'Hare in mid-1958. These flights touched down at Midway in just 16 minutes and reached O'Hare's Terminal 2 rooftop 20 minutes later.

Gary's fourth operator, Time Airlines, began service in 1969, putting the community on its route from Chicago to Detroit and Port Huron, MI. This airline soon folded, however, while traffic on the helicopter service remained weak. Whereas authorities had expected 40 to 50 people per day to use the helicopters, the actual number was closer to 12, according to the *Chicago Tribune*. This air-taxi service limped along for more than a decade, but by 1972 the helicopters had ceased stopping at Gary. A more traditional carrier, Rynes Airlines, tried to fill the void by flying small Twin Beech aircraft to O'Hare starting in mid-1975, but suspended service after a few months.[68]

For the next 14 years, Gary was back where it started, lacking any scheduled service and overshadowed by the region's larger airports.[69] The city's prospects were clouded by the plight of its economy, massive job losses in the steel industry, and the demographic specter of "white flight." Adding to its problems, the airport in Valparaiso gained traction and ultimately replaced Gary as northwestern Indiana's preferred point to access air-taxi flights. But Richard G. Hatcher, Gary's mayor from 1968 to 1987, continued to view the city's airport as a bright spot in an otherwise dire economy.

Prospects for the airport did not significantly improve until the municipal governments of Chicago and Gary signed the Chicago Gary Airport Compact in 1995, long after Mayor Hatcher had left office. After the agreement was signed, the airport was aspiringly renamed Gary/Chicago International Airport and investments were made in an improved terminal, an expanded parking lot, and other enhancements. As a new chapter for the airport began, a governmental authority created by this agreement offered significant financial incentives to entice prospective operators. Some cynical observers considered the advertising support and free parking provided to customers to be part of a grand strategy by Chicago Mayor Richard M. Daley to ensure that Gary

This *Chicago Daily Tribune* headline appeared on March 1, 1953, announcing the start of the first scheduled airline passenger service, provided by Lake Central Airlines, from Gary. (Author's Collection)

The bold lettering on the roadway sign at Gary/Chicago International Airport emphasizes its dual role serving the Gary area and its urban neighbor. (Author's collection)

eclipsed Peotone as the most likely site for the anticipated "third major airport" in the region.

The momentum continued after Pan Am Airways, a small start-up jet operator—unrelated to the famous, and defunct, global carrier of the same name—launched service to Las Vegas in 1999. Within a few years, the troubling pattern of airlines starting and stopping service reemerged—an embarrassment to the airport's promoters. The "new" Pan Am suspended service in 2002. Hooters Air and Southeast Airlines filled the gap in 2004 with service to the Carolinas and Florida, respectively, yet both were grounded by 2006. Rumors that JetBlue would launch its Chicago service from Gary circulated that same year, but the airline chose instead to operate from O'Hare.

Airport officials in Gary were understandably elated when the first Skybus Airlines jet took off for Greensboro, NC, in 2006. This ultra-discount carrier was soon providing seasonal service to four warm-weather destinations, offering as many as five flights from the airport on certain days. However, another dry spell began when Skybus entered bankruptcy and suspended service in 2008.

The main entrance at Gary/Chicago International Airport as it appeared in July 2013. A proposed expansion to the terminal is intended to increase the airport's appeal to major airlines. (Author's photo)

Almost four years passed before still another carrier, Allegiant Airlines, began service to Florida in early 2012. This airline operated two weekly round-trips to both Orlando and St. Petersburg, FL, but its tenancy played out the same way as its predecessors had, and it dropped Gary from its network in the summer of 2013.

Major improvements to the airport are again under way. With federal and state financial assistance, the main runway is being lengthened to support cargo and passenger operations with larger and more economical aircraft. The airport is also home to the Boeing Company's fleet of business jets serving Midwestern clients and has significant potential as an air cargo hub. Plans are also being made to expand its terminal. Many observers, however, question whether the airport will become more than a niche player in scheduled passenger service anytime soon. ■

Pan American Airways jets are wingtip to wingtip at the Gary/Chicago Airport terminal in 2001. This discount carrier, having purchased rights to the legendary Pan American World Airways name and trademarks, used mostly Boeing 727 aircraft such as these aging examples. This low-cost reincarnation of Pan Am suspended service to Gary the following year. (Photo provided by Robert Gyurko)

MEIGS FIELD
1400 S. Linn White Dr.

A Beechcraft B99 Airliner awaits passengers in front of the airport's terminal on June 9, 1975. While tiny compared to most aircraft used at Midway and O'Hare, this was considered a midsize airplane at Meigs. The terminal behind this Skystream Airlines "feederliner" boasted a soaring ceiling and offered splendid views of nearby Lake Michigan. (I.E. Quastler photo)

At its closure in 2003, Meigs Field left behind a legacy disproportionate to its size. This tiny lakeside airport, despite having only a single runway capable of handling only relatively small aircraft, hosted scheduled passenger flights almost continuously for more than a half century. The fate of Meigs came down to the relative merits of an airport versus an improved lakefront and museum complex. In the end, the City of Chicago favored the latter.

Hopes for a bona fide airport in downtown Chicago went unfulfilled for decades after air-mail service began on a landing strip in Grant Park in 1918. Both the city and state passed resolutions in the 1930s to create an airport near downtown, but this was not followed by immediate action.

The impetus for construction finally came in the early 1940s, as congestion at Municipal (later Midway) Airport grew steadily worse. Work on the modestly named "Northerly Island Landing Strip," built on the man-made peninsula called Northerly Island, began shortly after World War II. The new facility was opened for traffic in late 1946 and was dedicated in June 1950 in honor of Merrill C. Meigs, publisher of the *Chicago Herald* and *Examiner* and a strong supporter of aviation.

Scheduled passenger service began with air-taxi flights to Midway by the aptly named Midway Air Lines in late 1951. More substantial service began when

TAG Airlines made Meigs a stop on its Rockford–Detroit route and Lake Air Taxi began nonstop flights to Benton Harbor, MI, several years later. In 1957, Chicago Helicopter Airways launched service from Meigs with seven-seat Sikorsky S-55 helicopters shuttling passengers to and from Midway in a mere seven minutes, and to O'Hare in another 11 minutes. Within three years, CHA was flying nonstop 15 times daily to both Midway and O'Hare.

Meigs's small wooden terminal was replaced in 1961 with a modern glass-and-steel structure replete with office space, several ticket counters, and a spacious waiting area with splendid views of the lake. The mix of carriers serving the airport changed almost continuously, but Commuter Airlines, which flew to Cleveland, Detroit, Madison, WI, and Moline, Peoria, and Springfield, IL, was a stalwart. This carrier's Springfield flights were particularly popular, allowing lobbyists, legislators, and business travelers to make day trips, and even half-day trips, albeit mostly on aircraft with eight or fewer passenger seats.

Meigs Field came to be regarded as a preferred airport for business travelers making short-hop trips to and from downtown Chicago. By June 1968, *ten* scheduled carriers serving Meigs—Commuter, Gopher, Hub, Hulman, Midwest Commuter, Miller, Ong, Time, Trans Michigan, and Skystream—operated 122 scheduled weekday flights. They not only gave this tiny airport far more flights than once-mighty Midway Airport, which had only about 80 at the time, but more than any airport within convenient walking distance of the downtown of a major U.S. city has ever had to date.

The tarmac in front of Meigs Field's lakefront terminal is filled with aircraft, circa 1972. Meigs had extensive scheduled service at the time, although the number of flights had subsided somewhat since the 1968 peak. Construction of the Standard Oil Building is under way in the distance (beyond the terminal building, at left) while the John Hancock Center is newly complete. (Chicago History Museum, ICHIi- 51575)

MEIGS FIELD
1400 S. Linn White Dr.

The largest type of plane ever used in scheduled service from Meigs Field, this turbo-prop Hawker-Siddeley 748 is undergoing boarding for a trip to Springfield, IL, on June 9, 1975. As operated by Air Illinois, these 48-seaters were a popular way to reach the state capital for lobbyists, high-level governmental appointees, and members of the General Assembly. Lake Michigan can be seen in the background. (I.E. Quastler photo)

Meigs's services remained in almost constant flux over the next decade. Several airlines using the airport at its 1968 peak, including Commuter, went out of business. Chicago & Southern Airlines began service, only to have a flight originating at Meigs fatally crash in Peoria due to pilot error on October 28, 1971. This notorious accident, killing 14, sent shock waves through the commuter airline industry and led to heightened state regulation for these operators.

The next phase of air-service development stands out for the battle for market share to Springfield—the airport's most lucrative destination. In early 1972, Air Illinois expanded from five to eight trips serving the state capital each weekday, with continuing service to Carbondale, IL. Ozark Airlines grew concerned that this service was encroaching on its O'Hare business and, in March 1972, introduced nine weekday roundtrips on the same route. This gave Springfield-bound passengers a choice of 17 weekday arrivals and departures, and expanded weekend service as well. The future seemed to belong to Ozark, which was the larger and better-known carrier.

Such extensive service could not be sustained as oil prices rose and the economy weakened, and in a matter of months the number of round-trips plummeted. Ozark lasted less than a year before suspending all of its flights. In late 1973, Air Illinois reduced its schedule to five weekday round trips while at the same time introducing a 48-seat aircraft that reduced flight time to Springfield from 60 minutes to 45 minutes.[70] Larger and far more comfortable than the "puddle jumpers" they replaced, these planes raised hopes that a more diverse clientele would be attracted to Meigs.

But even these pared-down schedules could not be operated profitably. Helicopter service disappeared in 1976.[71] Then, on October 11, 1983, an Air Illinois flight that originated at Meigs crashed near Pinckneyville, IL, on its way to Carbondale, killing all 10 on board. After being grounded by the federal government, the airline resumed service to Springfield the following year. It limped along until it was replaced by Great Lakes Aviation, which concentrated on the Springfield route.

Efforts to close the airport intensified after its 50-year lease from the Chicago Park District expired in 1996. The city announced plans to close Meigs to support an ambitious lakefront improvement project, and greatly increased the fees charged to airlines for using its runways. As the shutdown loomed in August 2001, Great Lakes suspended its flights, citing its inability to compete with new service set to begin from Midway.

The future of the airport momentarily brightened when Mayor Richard M. Daley announced a compromise with the state that would allow the airport to remain open for another 25 years. But, in his view, the conditions upon which the deal rested were not met. On March 31, 2003, in the middle of the night, he had the runway abruptly removed from service. Many Chicagoans felt that the mayor neglected to obtain the necessary approvals and was "going too far" to achieve his lakefront goals.

Grassroots efforts to reopen the airport, however, proved no match against the powerful city administration. The FAA required the city to pay a fine to recover the federal funds that had helped improve the airport, but Meigs was gone forever as a functional airport. The terminal building and control tower were both preserved, with the former becoming a visitor center to support the lakefront park created on the site. ∎

Commuter Airlines timetable, March 1, 1966 (I.E. Quastler collection).

A pink limousine used by Skystream Airlines waits alongside Burnham Harbor, ready to shuttle passengers to wherever they want to go in the Loop area—one of perks of flying with this commuter airline. This photograph was taken in 1975, when Skystream was a major player at Meigs. (I.E. Quastler photo)

MEIGS FIELD
1400 S. Linn White Dr.

This map issued by Chicago Helicopter Airways for its January 1, 1960, timetable shows its assigned departure spot outside the Meigs Field terminal. The proximity of notable downtown hotels and tourist attractions is also noted. (Author's collection)

Shoehorned onto Northerly Island, Meigs Field was within walking distance of Soldier Field and the cultural institutions that today make up the city's famed Museum Campus (bottom and lower right). It was also only a short bus or taxi ride to McCormick Place (lower left) and the Loop. This photo shows the airport in 2003, shortly after the airport's decommissioning following a controversial decision by the city to replace it with a lakeside park. (Lawrence Okrent photo)

GENERAL MITCHELL INTERNATIONAL AIRPORT
5300 S. Howell Ave., Milwaukee, WI

The Layton Ave. Terminals, built in 1940, had notable architectural qualities, including slender vertical window panels next to the main entrance. The Works Progress Administration structure, however, might have been easily mistaken in this photo, circa 1949, for a government office building were it not for the sign above the entrance, control tower, and antenna masts on its roof. (General Mitchell International Airport)

For residents living on the northern periphery of the Chicago region, Milwaukee's General Mitchell International Airport is a heavily used alternative to O'Hare. Having long been a hub for discount carriers, Mitchell's importance to the debate about the region's airport needs has grown sharply in recent years.

Established as a private airfield in 1920 and sold to Milwaukee County in 1926, this municipally owned airport saw its first scheduled service in 1929. By the end of that year, 10 daily flights operated nonstop to five destinations: Chicago; Muskegon, MI; as well as Madison, Fond du Lac, and Oshkosh, WI.[72] In the airport's early years, a converted farm house assumed the unlikely role of terminal building.

By 1941, the airport was equipped with a conventional two-story passenger and freight facility, christened the Lawton Avenue Terminal, although the number of flights had grown to just 12. Passengers could only reach five destinations nonstop, but one of those endpoints—St. Paul, MN—provided attractive connecting opportunities for travelers headed to the West. Milwaukee's airport was renamed after the U.S. military aviation pioneer General William (Billy) Mitchell, in 1941.

By 1950, seven airlines operated 28 daily flights with nonstop service to nine cities. Northwest Airlines, a leader in serving the upper Midwest, gradually strengthened its dominant position. The airport's two-story terminal was increasingly inadequate, and on July 19, 1955, a new spacious three-concourse terminal with two levels and 23 aircraft gates was opened to the public. Despite these improvements, however, General Mitchell Airport was still greatly overshadowed by Midway and O'Hare. Milwaukee's airport did not yet have any nonstop service to points farther west than Denver or farther south than Washington, D.C. Jet operations did not begin until 1961, more than two years after the first U.S. domestic jet airliners entered service in December 1958.

Travelers eager for nonstop service on many long-haul routes had to wait until deregulation of the airline industry in 1978. Within two years, some of the gaps had been filled, and airlines flew nonstop from Milwaukee to 40 destinations. America West began nonstop departures to Phoenix, and Northwest Airlines to Los Angeles. But there were still major gaps in service, with nonstop flights to anywhere in Florida, Texas, and the Pacific Northwest still lacking.

Mitchell finally began to blossom in 1984 with the inception of Midwest Express Airlines. Touting itself as having "The Best Care in the Air," this enterprising operator offered first-class service at regular coach fares, featuring hearty meals, leather seats, and generous legroom. As a result, Midwest Express won numerous awards for its excellent service.

Two AirTran Airways Boeing 717-200s are at General Mitchell Airport in June 2013. AirTran attempted to purchase Midwest Airlines—the airport's dominant carrier—in 2006 and was acquired by Southwest Airlines several years later. Milwaukee remains one of its focus cities. (Author's photo)

GENERAL MITCHELL INTERNATIONAL AIRPORT
5300 S. Howell Ave., Milwaukee, WI

Milwaukee County Airport's original terminal, the converted Hirschbuehl farmhouse, was typical of many first-generation airport terminals. Small ground-floor windows, one of which was protected by an awning, and a gabled second floor used as the control tower provided less-than-ideal visibility for airline operations. This may have been the grand opening in 1927. (General Mitchell International Airport)

At first, Midwest only flew from Milwaukee to Atlanta, Boston, Dallas, and Appleton, WI, but by 1987 it had grown spectacularly, operating a large hub at Mitchell with nonstop service to more than 20 destinations, while its commuter airline partner, Skyway Airlines, flew to two dozen more cities. The carrier's expanding network and high-quality service, together with major renovations to the terminals, made Milwaukee the envy of many cities of similar size. Midwest's amenities—coupled with convenient parking and short walks to the terminal—elevated Mitchell's profile among travelers wary of O'Hare. In 2000, the airport had nonstop service to appreciably more cities (46) than even the resurgent Midway Airport (38).

Midwest Express, like all airlines, suffered greatly during the downturn following the terrorist acts of 2001. Another discounter, AirTran, launched Milwaukee service the following year, thereby making Midwest's situation even more precarious. After changing its name to Midwest Airlines, the carrier trimmed expenses and eliminated some of its in-flight service, including its complimentary meals. These efforts fell short, however, and the cash-strapped carrier was sold to Republic Airlines in 2009, becoming part of Republic subsidiary Frontier Airlines. Within a few months, Midwest's former Milwaukee hub had been dismantled.

The downturn in service was relatively brief. Southwest Airlines arrived to fill the void in late 2009. Instead of driving to Mitchell for the in-flight meals and added legroom, passengers increasingly made the trip to save money and take advantage of Southwest's flexible ticketing policies. Southwest also purchased its former rival AirTran, thereby further increasing its local dominance. As the reconstruction of the Tri-State Tollway aggravated traffic congestion around O'Hare in 2010, Mitchell's popularity soared, particularly among residents living in Chicago's northern suburbs. In contrast to the drive to Midway, which could be as long as two hours, some of these northern suburbanites could reach Mitchell, on Milwaukee's south side, in 45 minutes or less.

Growing traffic at the Amtrak station next to Mitchell's main parking lot, which had opened in 2005, further raised the airport's visibility.[73] Passenger-train advocates, however, wanted much faster and more extensive service, and touted the potential of high-speed rail between Milwaukee and Chicago, with stops at both Mitchell and O'Hare. Wisconsin officials reportedly balked at this idea, however, due to the associated costs and concerns that such a plan might boost O'Hare at the expense of Mitchell.

As pressure grew to create a third major airport for the Chicago region, Mitchell assumed the role of spoiler. While detached from the planning process of its giant neighbor, Mitchell diverted enough business from the Chicago region to raise questions about whether an additional airport was a viable proposition. While not going so far as adding "Chicago" to its name, as Gary and Rockford have done, Mitchell is aggressively marketed as "conveniently located in Milwaukee to serve southeastern Wisconsin and northern Illinois, including Chicago."

Passenger traffic at Mitchell is still well below levels of the all-time peak in 2010, when Midwest Airlines shut down. Even so, Southwest remains on an upward trajectory, and the concerted push to attract flyers from the Chicago market appears poised to grow even stronger. Travelers at Mitchell will find a historic gallery maintained by the nonprofit Museum of Flight recounting the many phases of this airport's past. ■

Passengers amble through the glass-topped Concession Mall, an area notable for being accessible to travelers and non-travelers alike, at General Mitchell Airport. The old-fashioned clocks hanging in the center of the waiting room were custom built to resemble those common in Milwaukee in the late 1800s. (Mke-airport-terminal, by Americasroof, is licensed by CC BY-SA 2.5)

General Mitchell International Airport, shown in this photograph following the completion of a major expansion project in 1986, featured a new control tower, large parking garage, and three lengthy concourses. The 1955 terminal was largely demolished during this ambitious expansion. Traffic grew so rapidly over the next several years that in 1990 a perpendicular addition was made to the middle concourse. This addition gave it the shape of a hammerhead, and hence it is often called the "D Concourse hammerhead." A similar addition was later added to the C Concourse (upper right). (General Mitchell International Airport)

A Skyway Airlines commuter jet is parked at General Mitchell Airport on January 7, 2007. This Fairchild-Dornier 328, operated under contract by Midwest Airlines, was assigned to "Midwest Connect" feeder routes to smaller cities. (Mke-airport, by Americasroof, is licensed by CC BY-SA 2.5)

CHICAGO ROCKFORD INTERNATIONAL AIRPORT
2 Airport Cir., Rockford, IL

Chicago Rockford International Airport's present terminal, built in 1987, features a spacious ticketing hall with high ceilings and secure facilities for international flights. Building such a large terminal to attract new airlines was at first unsuccessful, but the gradual expansion of passenger flights since 2005 has generated considerable optimism about its market potential. (Author's photo)

Like many other airfields a little more than an hour's drive from a major hub airport, Chicago Rockford International Airport has faced a difficult up-and-down pattern of scheduled air service over the past sixty years. Once a facility that predominantly served business passengers traveling to and from a geographic area with a relatively small population, its scheduled service today serves mostly pleasure travelers from a broader region. Recently, this airport's nonstop service to popular Sunbelt destinations has elevated its importance to those living in the northwest suburbs of Chicago.

Created to support a U.S. Cavalry detachment in 1917, this airfield—initially called Camp Grant Airport—was used primarily for military purposes for nearly 30 years. The creation of a municipal airport authority in 1946, however, was a springboard for change, resulting in major terminal improvements and lengthening of the main runway. At the newly renamed Greater Rockford Airport, scheduled service began with Mid-Continent Airlines flights to Midway Airport and several Iowa points in September 1950. Service continued after Mid-Continent merged with Braniff Airlines in 1952.

The airport soon attracted several other commercial airlines. Ozark Air Lines launched Rockford–St. Louis service in 1951 and soon also flew to the Twin Cities, Omaha, and other points north and west of Rockford. Lake Central Airlines began stopping its Chicago–Madison, flights at Rockford in 1953. By 1955, Ozark had taken over the last of Braniff's routes and become Rockford's dominant carrier.

Within a few years, Rockford encountered a problem that would hamper it for decades: its close proximity to airports in Chicago undermined consumer interest in its flights. As airlines transferred flights from Midway to the much closer O'Hare Airport in the late 1950s, this problem grew worse. The new airport was a mere 75 miles away and could be reached by automobile in just 75 minutes after the Northwest Tollway opened in 1958. Two years later, a popular bus service began between Rockford and O'Hare.[74]

Despite the ease of driving to airports in Chicago, North Central Airlines began having its Chicago–Madison flights make stops in 1959. Gradually, however, the momentum was lost. By 1970 Rockford's nonstop service had been reduced to flights to Dubuque, IA, and O'Hare.

However, hopes that Rockford's scheduled service would once again rise to a position of prominence returned several years later. Ozark launched Rockford–

Guests at the Rockford Air Show stand in line to see the inside of a rare visitor to the city—an American Airlines Boeing 707, circa 1968. Hopes were high at the time that Rockford Airport would soon gain jet service to high-profile destinations. Within a few years, nonstop jet flights were introduced to New York and Washington, D.C. Unfortunately, these services proved short-lived. (Chicago Rockford International Airport)

Passengers wait in the departure lounge at Chicago Rockford for an Allegiant Air flight to Punta Gorda, FL, on June 29, 2013. Several other Allegiant flights departed this same day, keeping the lounge relatively busy. (Author's collection)

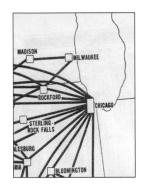

This Ozark Airlines map, appearing in a 1957 company timetable, shows four routes serving Rockford and more than a dozen reaching Chicago. For years, the St. Louis-based carrier's frequent service to Midway Airport was a Rockford mainstay, although the ease of driving to the Windy City's airports had always weakened demand. (Author's collection)

Denver service, with a stop in Iowa, in 1975. More good fortune came to Rockford in 1978, when the mercurial Coleman Air Service made Rockford a hub to an entire network of routes using Jetstream I airplanes with a dozen seats. Coleman began flying to Cedar Rapids, Cincinnati, Detroit, Des Moines, Indianapolis, Peoria, and other points. To enhance its fledgling hub, it also began flying DC-9s between Rockford and New York City's LaGuardia Airport, pushing the community's air service into an entirely new realm. By 1979, Rockford had nonstop service to more than a dozen cities. In barely a decade, Rockford had gone from an airport with only "puddle jumpers" to one having an impressive array of long-haul flights.

TWA stepped into the fray with flights to St. Louis, but this golden era of nonstop service could not be sustained. Coleman lasted only a few years and abandoned all but the Detroit flights (which lasted only a few more months) by early 1980. Ozark dropped its Denver departures and TWA pulled out at approximately the same time, reducing Rockford's service to just two carriers, flying predominantly to Minneapolis and O'Hare.

Rockford found itself once again a lightly served airport in the shadow of Chicago. Rather than offering flights to Rockford, United Airlines established a ticketing arrangement that allowed travelers to fly into O'Hare and finish their journeys on buses. A new passenger terminal built in 1987, while impressive for its size, did little to stimulate new business. Instead, American Airlines followed United's example and replaced its Rockford flights with bus connections.

The airport's cargo-handling business, meanwhile, performed spectacularly. United Parcel Service made Rockford an important hub and package-sorting center in 1993. Even as UPS expanded, however, scheduled passenger service remained on a downward trajectory. By 2000, the airport was back to having nonstop flights to just Detroit and Dubuque, IA. Even these limited offerings languished. After the exit of Northwest Airlines in 2001, Rockford joined the growing list of U.S. airports that had lost all scheduled passenger service.

Rockford regained a foothold in the passenger market when Trans Meridian Airlines began flights to Florida in late 2003. Northwest and Hooters Air soon followed suit, but all were gone by early 2006. Fortunately, the void was filled that same year with twice-weekly Allegiant Air flights to St. Petersburg, FL, and those of another discounter, Apple Vacations, flying to Cancun, Mexico. When Frontier Airlines launched thrice-weekly service to Denver in late 2011, allowing for convenient connections to the West Coast, and Allegiant began flying to Las Vegas and other leisure spots, albeit with similarly limited frequency, Rockford once regained a sense of momentum.

The airport was renamed Chicago Rockford International Airport in 2005 to emphasize its relevance to flyers in the Windy City. While it is not seen as a potential "third airport" for the region, Rockford has been able to attract the interest of scheduled airlines in ways that neither Gary/Chicago International Airport nor the proposed Peotone airport have. (Some of the airport's flights are made possible by local "risk mitigation" agreements that guarantee the airline a certain level of passenger revenue). Gradually, it has become a more prominent player in the policy debate about the future of air transportation in metropolitan Chicago. The expansion of cargo operations affords it financing options that most other small airports lack and gives it the opportunity to aggressively promote itself to the vast Chicago regional market. ∎

Chicago Rockford International Airport's logo—depicting a large, stylized airliner above a wide runway—incorporates the facility's three-letter identification code, RFD.

Passengers having just arrived from Florida wait outside a rental car counter at Chicago Rockford International Airport in 2013. (Author's photo)

SOUTH BEND REGIONAL AIRPORT
4477 Progress Dr., South Bend, IN

South Bend Regional Airport, while buffeted by sharp swings in airline service for decades, has consistently attracted business from the eastern edge of the Chicago metropolitan region. The airport's success in winning this business casts a shadow over efforts to reestablish scheduled passenger service from closer cities, particularly Gary/Chicago Airport.

Scheduled service to South Bend began within months after inventor Vincent Bendix purchased land northwest of the city for an airport in 1929. His timing was fortuitous: Stout Air Services launched its pioneering flights between Chicago and Detroit—one of the country's first scheduled air services—that same year. Each of Stout's four daily flights in each direction, provided with three-engine Ford Tri-Motors, made stops. These flights combined with revenues from Bendix Corp., a major producer of braking systems for automobiles, made major improvements to the airfield possible. It was renamed Bendix Municipal Airport in 1932.

The level of service available from South Bend's airport, however, did not expand as rapidly as might have been expected at the time. In 1940, the airport had only five daily flights, and nonstop flights to the same three destinations as in 1929—Battle Creek, MI, Chicago, and Detroit. Direct service had become available to Boston and New York, but reaching these cities required making several stops.

The 1949 opening of a strikingly beautiful new terminal, designed in the International Style and notable for its extensive use of glass, was a turning point. The facility was renamed St. Joseph County Airport and by 1955 offered nonstop service to six cities and one-stop to another five. When the Indiana Toll Road opened in 1956, tens

of thousands of residents in northwestern Indiana could reach this airport in about an hour—far less time than it took them to reach O'Hare. Eager to tap into the growing business-travel market, United Airlines in the late 1950s commenced one-stop South Bend–Newark, NJ, flights via Fort Wayne (later via Cleveland). By 1960, airlines flew nonstop from South Bend to eight cities.

Equally exciting was the ability to overfly Chicago on trips to South Bend from the West, which became possible when United began nonstop flights from Denver (which continued east to Fort Wayne) in 1976. South Bend, now an up-and-coming airport, began attracting the attention of many prospective flyers.

Officials had reason for optimism in 1974 when the facility was renamed Michiana Regional Airport to reflect its increasing importance to the Michigan–Indiana region. Although the flights to Newark were dropped in 1978, the airport took on the aura of a big-city facility with the 1982 opening of a new terminal building with more gates, enhanced dining and retail space, and improved facilities for intercity buses. United ended its Denver flights in

A North Central DC-9, en route to Chicago or Detroit, is being readied near the control tower at South Bend circa 1975. Loaded baggage carts near the nose suggest that boarding has already begun. The rooftop railings above the terminal building are part of the observation deck—a popular amenity for visitors to airports at the time. (South Bend Regional Airport)

1987, but Delta and USAir (later renamed US Airways) partially filled the void with nonstop service to Cincinnati and Pittsburgh, respectively. In November 1992, the airport became the terminus for South Shore Line trains, making the airport a point of embarkation for intercity travelers on buses, planes, and trains—a distinction that no other Midwestern city shared at the time. These trains (which were only nominally "intercity," running only between Chicago and South Bend) had previously ended their runs near the local Bendix plant.

By the late 1990s, however, the airport's vulnerability was again exposed due to a surge in discount flights offered from Chicago's Midway Airport and the growing unwillingness of passengers to make time-consuming connections. By 2000, South Bend lacked service to points outside the Midwest. The subsequent introduction of Delta flights to Atlanta raised hopes for a turnaround, but this was not to be. Flights to Pittsburgh were eliminated by 2003, and the Atlanta service lasted only a few more years, leaving only flights on small aircraft operating to regional hubs at relatively high fares. South Bend's once-vibrant airport was in a deep slump.

A regional jet flown by Comair, operating as Delta Connection Flight 5748, arrives in the South Bend Airport from Cincinnati on August 1, 2004. This photograph was taken from the airport's viewing deck—a favorite among airline enthusiasts. (Charles Juszczak photo)

In 2003, officials renamed the facility yet again, christening it South Bend Regional Airport, to better reflect its location and strong connection to its host city. A modest recovery in service also began. Not only did the flights to Atlanta return, but Allegiant Air began twice-weekly nonstop service to Las Vegas (the airport's longest-scheduled flight since its inception) and Orlando-Sanford

Airport in 2005. Gradually, the airport clawed its way back, and today, it has 24 daily flights operating nonstop service to six cities.

South Bend Regional Airport's heightened role to residents of northwestern Indiana today makes it increasingly important to metropolitan Chicago's air-travel market. The airport, however, is not widely considered a major contender to provide additional airport capacity in the Chicago region. Although situated on the northwest side of South Bend, it remains too far east of Chicago to be prominent in discussions about becoming the region's third major airport. ■

Passengers walk down the high-level platform at the South Shore Line station at South Bend airport on January 18, 2006. The South Shore Line, like Coach USA and Greyhound, departs from boarding areas that open directly to the main ticketing area, making this airport a case study for intermodalism. (South Bend South Shore Line, Georgi Banchev, is licensed by CC BY-SA 3.0)

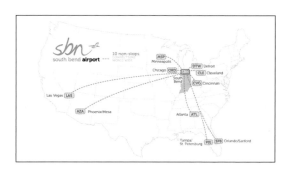

A South Bend Regional Airport map touts the city's expanding route network. The long-haul flights to Florida and the Southwest operate several times weekly rather than daily. (South Bend Regional Airport)

AIR-TAXI SHUTTLE DESTINATIONS
Terminal Town

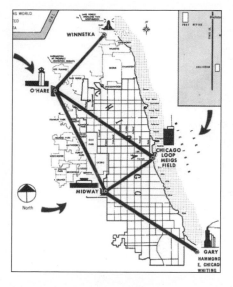

Chicago Helicopter Airways was at its zenith when this map was printed in its January 1, 1960, timetable. The helicopter operator's system of routes linked Winnetka, Gary, and Meigs Field to both Midway and O'Hare. (Author's collection)

Travel times between many points in the metropolitan Chicago region are great enough that air-taxi service to Midway and O'Hare airports has had considerable practical value. Some airports have had on-demand service (similar to that of taxicabs carrying customers who made reservations in advance), while others have had flights operating on advertised schedules, much like major airlines. Most air-taxi providers used small, propeller-driven aircraft, but helicopters were also at times part of the mix.

Precisely when the region's first air-taxi service began is unclear, but it appears that *scheduled* service (available to the general public at designated times between advertised points) began with United Airlines flights linking Glenview, IL, and Municipal (Midway) Airport on April 1, 1938.[75] Far more significant to the region, however, was "the nation's first intra-city air shuttle service" launched in 1951 between Sky Harbor Airport in Northbrook, IL, and Midway Airport. Within a year, Midway Air Lines, the operator, also touched down at DuPage Airport (West Chicago) and Meigs Field. Several years later, a competitor began a similar shuttle service to Midway from Chicagoland Airport near Wheeling, IL.

Air-taxi service was particularly popular with business travelers wanting to avoid stressful and time-consuming drives to the airport. Prior to the opening of the region's expressways, motorists from some suburbs could expect to spend up to two and a half hours behind the wheel due to frightful congestion, low speed limits, and frequent stops due to the prevalence of traffic signals. With aviation fuel and landing fees at airports less costly than they are today, scheduled air-taxi service could often be profitable with only two or three paying passengers per flight. In some cases, major airlines picked up the tab for connecting passengers buying first-class tickets.

(Opposite) The airports shown in red on the map at left were points of origin for scheduled air-taxi flights to Midway and O'Hare. All of these locations except the Winnetka Heliport, which was exclusively used for helicopter service, once had a half-dozen or more daily "fixed wing" (conventional airplane) departures for these busy hubs. Meigs Field, which had both intercity flights and air-taxi service for sustained periods, is discussed in the chapter devoted to commercial airports.

AIR-TAXI SHUTTLE DESTINATIONS
Terminal Town

The region's air-taxi service reached its zenith when scheduled helicopter service to Chicago's two major airports began in 1956. All air-taxi operators, however, suffered a powerful one-two punch in the form of the gradual relocation of flights from Midway to O'Hare and the completion of new expressway and tollway systems. With the exception of helicopter flights between the city's airports, what little remained of the region's air-taxi service vanished as travelers became acclimated to driving on newly opened superhighways after 1958.

A second phase of air-taxi service began approximately a decade later to serve travelers in northwestern Indiana who had been inconvenienced by the relocation of flights to O'Hare. By the early 1970s, scheduled air-taxi service was available to O'Hare from smaller airports in Gary, LaPorte, Michigan City and Valparaiso, IN.

The following pages feature the airports that were termini for scheduled air-taxi routes to Midway and O'Hare. Several others that were either intermediate stops, or were served only by airlines principally oriented toward hauling mail, are not included. (Please refer to Appendix II for a map of air-taxi routes since 1939). Today, time sensitive travelers weary of expressway congestion often yearn for the restoration of scheduled air-taxi service. At least for the moment, their desires appear to be at odds with market realities. ■

This postcard of Chicago Helicopter Airways shows a flight above Grant Park, with the Prudential Building and Illinois Central yards in the distance, circa 1960. (Author's collection)

A Midway Airlines Cessna 19A takes off from Sky Harbor Airport in Northbrook, IL., on May 20, 1952. (Author's collection)

(Opposite) Airliners of many makes and models await departure from Midway Airport's gates, circa 1955. The small plane, a Midway Airlines Cessna 19A, made air-taxi flights to Meigs Field and the Chicago suburbs, including DuPage Airport in West Chicago and Sky Harbor Airport in Northbrook. The much larger Wisconsin Central Airlines DC-3 is likely destined for Madison or Milwaukee. (Pat Bukiri photo)

CHICAGOLAND AIRPORT
Half Day Rd. & U.S. Route 45, Lincolnshire, IL

An easterly wind blows as a small plane prepares to land on Chicagoland Airport's only paved runway on June 2, 1978. This photo, looking south from the vicinity of Half Day Rd., illustrates the large private aircraft fleet based at the Lincolnshire airport. Milwaukee Ave. can be seen in the distance at left. (Jim Frost photo, *Chicago Sun-Times*)

Ambitiously named and attractively situated in the midst of a rapidly growing suburbia, Chicagoland Airport seemed poised for an important role in scheduled passenger service in the mid-1950s. In the end, however, this facility was, like most small airports in the region, relegated to footnote status in that travel sector. Chicagoland never became more than the terminus of a short-lived shuttle service to Midway Airport.

When it opened in the early 1940s, the primary purpose of this small airport in the village of Half Day (today part of Lincolnshire, but near that community's boundary with Vernon Hills) was to support World War II pilot training at Glenview Naval Air Station. After the war, the airport suffered from downturn in military activity and was nearly abandoned.

The airport reopened as Chicagoland Airport in 1946 and experienced a surge in civilian flight activity, despite the absence of paved runways. Scheduled service began when the airport became a stop on Chicago Airways' Midway–Milwaukee route in 1955.[76] This short-hop service, consisting of three daily flights in each direction, also made a stop at newly opened O'Hare (listed in timetables as Park Ridge due to the tendency of air-taxi operators to name their stops for communities rather than airports) and appears to have been oriented heavily toward carrying mail. In 1957, Chicago Airways began advertising three-times-daily shuttle service to Midway in the *Official Airline Guide*. These flights operated in the early morning and late evening, apparently to accommodate passengers making connections. One-way fares were $10 (about $78 in today's dollars), and flights reached Midway in just 20 minutes.

Travelers flying to and from this airport experienced bumpy takeoffs and landings on grass runways. Air-taxi service appears to have ended in 1958, when advertisements in the *Official Airline Guide* ceased. Although Chicagoland continued to serve many private airplanes—and its main runway was eventually paved—encroaching residential development posed a threat, and the airport was closed in 1978.[77]

Less than a decade later, much of the site had given way to industrial development and new streets, including aptly named Schelter Rd.—named after Art Schelter, who was the airport's owner for many years. The straightening of Half Day Rd. obliterated some of the airport's remaining structures.

Visitors looking for remnants of this "ghost airport" will find scattered concrete remnants near the intersection of Half Day Rd. and Milwaukee Ave. Some crumbling vestiges immediately southwest of that intersection are said to be part of the foundation of the airport's beacon, used by pilots for navigation.■

CHICAGO AIRWAYS
PASSENGERS AND FREIGHT

Chicago Airways, the only scheduled carrier ever to serve Chicagoland Airport, had a letterhead that emphasized its reliance on both passenger and freight service. (Author's collection)

A photograph taken in 2014, from almost the same location as the 1978 view, shows vacant land on the former site of Chicagoland Airport. Hardly a trace of the runway remains. (Author's photo)

DuPAGE AIRPORT
2700 International Dr., West Chicago, IL

The relatively dense development surrounding DuPage Airport is evident in this April 22, 2012, photograph. While heavily agricultural into the 1940s, this area saw a surge in construction before air-taxi service began in 1952. The original short runways were configured in an "X"; remnants, visible at left, seem a mere afterthought today. (Chicago - DuPage Airport (KDPA - DPA), by Magic Aviation, is licensed by CC BY-SA 2.0)

DuPage Airport has a notable distinction in the region's airline history: its 32 scheduled daily passenger flights in 1956 were the most ever offered at a suburban airport from the dawn of commercial aviation to the present day.[78] Most of these flights were shuttle trips to and from Midway, taking only 17 minutes.

For many years after this west-suburban airport opened in the 1920s, it had only grass runways and appeared to have little prospect of attracting a scheduled airline. After being requisitioned by the U.S. Navy in 1941, its prospects improved. Two runways were paved and several large hangars were built. Hundreds of airplanes manufactured on the property by the Simpson Aircraft Co. contributed to the bustle of wartime traffic. Still, scheduled airlines showed little inclination to add flights to the airport.

After the Navy turned the airport over to the DuPage County government in 1946, the airlines began to take notice. A boom in residential construction began to reshape the airport's rural surroundings. In June 1952, Midway Airlines received authorization to provide air-taxi service. The carrier's flights shuttled passengers on eight-seat Piper aircraft to Midway Airport. Midway Airlines advertised this suburban airport as "St. Charles" due its proximity to the prosperous Fox River community, which straddles both DuPage and Kane counties.

> **Okay Daily Flights From Chicago to DuPage Airport**

A *Bensenville News* headline on June 30, 1952, announced approval for Midway Airlines' service to DuPage County Airport. (Author's collection)

Midway Airlines' flight schedule expanded to 12 eastbound trips and 14 westbound trips, and in 1955 a second airline, Chicago Airways, launched a multiple-stop service to the airport that same year. DuPage Airport was the second stop (after Mitchell Field in Lombard) on Chicago Airways' flights from Midway to Rockford, IL. This carrier's three daily round-trips brought the number of scheduled DuPage County Airport passenger takeoffs and landings on peak days to 32.

It is not clear when these services ended. Both carriers ceased advertising in the *Official Airline Guide* in 1958, as flights were being rapidly shifted from Midway Airport to O'Hare. Despite almost continual modernization of DuPage Airport, scheduled passenger flights never returned, except for a short-lived Crescent Helicopter service that advertised trips to O'Hare in the early 1980s.[79] These flights, however, were apparently never operated on fixed schedules. Today, DuPage Airport is one of the busiest "reliever airports" (i.e., one relieving congestion at a commercial-service airport and serving general aviation) in Illinois and is home to more than 350 aircraft. ■

The management team at DuPage County Airport in West Chicago poses in front of a Piper Twin Comanche, circa 1962. The hangar sign heralds DuPage as "Chicago's Better Weather Airport"—a reputation attributable in part to the airport's ability to remain operational when fog or "lake-effect" snow cripples airports closer to Lake Michigan. (Courtesy of DuPage Airport)

GLENVIEW NAVAL AIR STATION
1951 Tower Dr., Glenview, IL

A U.S. Marine Corps detachment stands at attention in front of Glenview Naval Air Station's control tower on June 26, 1953. Scheduled air-taxi service—available to both civilians and military personnel—was provided at this airport into the early years of World War II. (Chicago History Museum; ICHi-37915)

Curtiss Field, renamed Naval Air Station Glenview during World War II, was a pioneer in the development of air-taxi service into the suburbs. Between 1938 and 1941, United Air Lines flew nonstop between this historic airfield and Municipal Airport in Chicago, allowing for convenient transfers to the East Coast and other destinations.

Air-taxi flights serving Glenview straddled the general aviation (civilian) and military phases of this airport's history. The general aviation phase began when Curtiss Field opened in 1929—and is remembered in aviation history for the Curtiss Aeroplane Company's popular "flying service" and heavily publicized airplane races featuring celebrity pilots such as Jimmy Doolittle, Charles Lindbergh, and Wiley Post. Curtiss Field was also once home to the largest airplane hangar of its time—the massive Hangar One.

This United Air Lines' "Coast to Coast" logo appeared on Boeing aircraft operating to Glenview in 1938. (Author's collection)

Hopes ran high that Curtiss Field (also called Curtiss-Wright Field) would become a major competitor to Municipal Airport. In 1933, the airport basked in the spotlight when the German airship Graf Zeppelin, landed in Glenview. The Zeppelin strategically circled over Chicago's Century of Progress exhibition, in order to hide its controversial swastika-emblazed tailfin. (Hoping that the Nazi swastika on massive airship's starboard side would not be seen by fairgoers, authorities permitted it to circle the fair only in clockwise fashion). The optimism of the early 1930s, however, faded when major improvements to the airport failed to attract any scheduled airlines and plans to build a modern terminal were scrapped.

The airport received a break in early 1938, when United Airlines announced a shuttle service from Glenview to Municipal Airport using twin-engined Boeing aircraft. This service, launched with considerable fanfare on April 1, 1938, allowed travelers to reach Municipal in less than 20 minutes. By sparing motorists the need to contend with severe highway congestion, these flights reportedly saved as much as two hours of travel time. Special sightseeing flights on summer weekends gave residents a sampling of the service, while the *Glenview View* newspaper heralded the sophistication of the airline's Boeing aircraft, which were capable of traveling three miles per minute. United publicized the fact that travelers could easily transfer at Municipal to flights bound for Cleveland, New York, Philadelphia, and other cities.[80] The trip from Glenview to New York, for example, took a mere 5 hours and 20 minutes.

Curtiss Field was sold to the U.S. Navy in 1940, beginning the second and final phase of the airport's history. As massive investments were made to ready the airport for military use, it was renamed Naval Air Station Glenview, and most civilian flights were phased out. United's operation was briefly moved to Northbrook's Sky Harbor Airport around 1941, and then discontinued.[81]

The Air Station—one of the three most important naval airports in the country—shouldered an enormous burden through World War II and into the postwar era, but prospects for the return of scheduled passenger service dimmed after O'Hare opened in 1955. The Navy's need for the airfield gradually declined, and by the late 1970s, the ongoing value of Glenview Naval Air Station was being questioned. The facility was identified for closure in 1993. After a lengthy decommissioning process, the last parcels of land were transferred to civilian ownership, with most of them in the hands of the village government and its specially created development authority.

The Glenview Naval Air Station control tower is now part of a large retail-oriented development. The structure is one of the few large-scale transportation landmarks in the region that was preserved after it ceased being used for its original purpose. (Xhoana Ahmeti photo)

Demolition crews removed more than one million cubic yards of concrete. Some 108 former military buildings were relocated or demolished. The enormous development of homes, offices, parks, and retail space that took their place—"the Glen"—was completed in 2003 and is widely considered to be a redevelopment success story. The preserved control tower, listed on the National Register of Historic Places, was turned over to the Glenview Hangar One Foundation. Most of Hangar One, however, was dismantled despite concerted preservation efforts by the foundation and the U.S. Navy. ■

PHILLIPS FIELD
1300 North Hwy., Michigan City, IN

The Phillips Air Taxi fleet is positioned on the tarmac at Michigan City, IN, circa 1970. The sign encourages customers to "Go with Joe!" Phillips offered air-taxi service to and from O'Hare Airport, a little more than 80 miles away. (Courtesy of Phillips family)

Phillips Field, opened by businessman Joe Phillips in the 1940s, had only a grass runway and seemed to have little potential for scheduled service. The gradual relocation of flights from Midway to O'Hare starting in the mid-1950s, however, created a new opportunity. Travelers in Michigan City faced the difficult prospect of driving more than 80 miles to reach Chicago's primary airport. As roads became more congested, a local air-taxi option became much more attractive.

After making improvements to his airport and installing a paved runway, Phillips launched a namesake airline in 1969. Making four weekday flights to O'Hare and back, his propeller-powered Piper aircraft (all with eight or fewer seats) covered the distance in just 30 minutes. Patronage appears to have been light, however. In 1971, the airline consolidated its Michigan City flights with a service from Porter County Airport in Valparaiso, IN, that had commenced at approximately the same time. This stop at the county seat added 10 minutes to the trip to O'Hare.

Competition escalated when bus service between Michigan City and O'Hare was introduced by Tri-State Coach Lines in 1973 from a terminal only a few miles

Located beside U.S. Route 20 (lower right), the Phillips terminal was a busy place circa 1970. Looking somewhat out of place, the large aircraft is a Convair CV-440 owned by the Chesapeake & Ohio Railway. It has brought executives in from the company's Cleveland headquarters to visit the South Shore Line, a railroad based in Michigan City and owned at the time by C&O. (Courtesy of Phillips family)

from the airfield. Tri-State's service was much less expensive and operated with greater frequency. In 1977, Phillips Airlines added yet another stop to its route, in nearby La Porte, IN, adding 10 more minutes to each flight.

The end came in late 1986, when Phillips Airlines suspended service due to insufficient demand. The Michigan City municipal government purchased Phillips Field that year and transformed it into a public airport. The small Phillips Airlines terminal building was torn down and replaced with a modern structure. A marker commemorates the pioneering role of Joe Phillips in the airfield's development. ■

This attractive terminal building at Michigan City Municipal Airport, which replaced the Phillips Air Taxi terminal on the opposite side of the airfield, has a plaque near its entrance commemorating the pioneering contributions of Joe Phillips. The entrepreneur sold his airport to the city after scheduled service ended in the 1980s. (Author's photo)

SKY HARBOR AIRPORT
500 Anthony Tr., Northbrook, IL

A Ford Tri-Motor belonging to Gray Goose Air Lines is parked in front of the Art Deco terminal at Sky Harbor Airport in Northbrook, IL, circa 1935. By the time air-taxi service began in 1952, this beautiful structure had already been torn down. (Northbrook Historical Society)

Northbrook's Sky Harbor Airport gave birth to the nation's first intra-city air-shuttle service.[82] The community's 15 daily scheduled passenger departures to Midway in 1955 were expected to be only the beginning of its expanding role in commercial aviation.[83]

Optimism about commercial air service from Chicago's north suburbs grew sharply in the mid-1920s, when the North Shore Airport Co., a division of United Aviation Corp., began work on this ultra-modern airport at the intersection of Dundee Rd. and Anthony Trail. The "Airport of Tomorrow" that opened in 1929 had several runways, a terminal built in the Art Deco style, and a large hangar. Area residents took delight in the elegant Petrushka Club atop the terminal, where patrons could dine, dance, and observe aircraft.

The timing of Sky Harbor's completion, however, could hardly have been worse. The Great Depression and competition from other airports hampered its financial performance. Short-lived Grey Goose Airlines offered air-taxi service, but it is unclear whether it was ever provided on a scheduled basis. By 1939, Sky Harbor had been abandoned, its terminal vacated, and its popular clubhouse vandalized. After being sold to several individuals, the terminal building was torn down and the airport took on a new role—that of a flight-training center for the U.S. Navy to support the nearby Glenview Naval Air Station.

Sky Harbor's new role was strengthened by the construction of a second hangar in 1941. Then United Airlines' scheduled air-taxi service from Chicago Municipal (later Midway) Airport that had previously operated to Glenview NAS began operating to Sky Harbor instead. Although this service apparently lasted no more than a few months, opportunity continued to beckon. By 1950, two of Sky Harbor's four runways had been paved. On June 29, 1951, Midway Air Lines (not the same company as the later jet operator) began a much more substantial air-taxi service to Midway Airport using single-engine Cessna 19A aircraft that were compatible with the airport's short runways.[84] The trip from Northbrook (the company headquarters) was scheduled for 17 minutes. No flights were operated at night for safety reasons.

Midway Airlines added flights from Midway to Meigs Field in the autumn of 1951 and soon expanded to DuPage Airport in West Chicago. Within a few years, the carrier, described as the "world's shortest airline," was carrying upwards of 2,000 passengers per month.[85] An interline agreement between United Air Lines (later renamed United Airlines) and the air-taxi provider allowed travel between any points served by the two carriers' systems on a single ticket.[86] The Northbrook-Midway route, however, was its flagship service and grew to 15 daily round trips.

The financial viability of these flights plummeted as commercial flight activity in the region shifted from Midway to O'Hare, which was much closer to Sky Harbor. Although the airline received permission to fly to O'Hare, the short distance between Northbrook and the new airport limited passenger demand. By 1958, all air-taxi service had ended, and two years later, the future of Sky Harbor was in doubt.

Noted developer Arthur Rubloff bought the property in 1968 with hopes of creating the first "air-industrial park" in this part of the country, attracting manufacturers who would maintain their own airline fleets. This, however, was not to be, and Sky Harbor was closed in 1973. After streets were built through the property, redevelopment eventually altered most of the airport site beyond recognition. Two of the hangars survive on the 500 block of Anthony Trail. ■

Midway Airline's logo featuring a luggage-toting kangaroo jumping over a city skyline emphasized the company's short-hop flights to and from Chicago's airports. (Author's collection)

This iconic hangar at Sky Harbor stands as a reminder of the high expectations for this Northbrook airport years ago. Today, the well-preserved building is used by a construction company. Another historic hangar is located just behind this graceful 1929 landmark. (Author's collection)

VALPARAISO (PORTER COUNTY REGIONAL) AIRPORT
4207 Murvihill Rd., Valparaiso, IN

Opened in 1949, this modest terminal building at Valparaiso's Porter County Airport, shown circa 1975, was a rarity for having a covered front porch (at far right). Passengers using Phillips Airlines could park and walk to the gate in less than a minute—or relax and enjoy the breeze on the porch. (Photo by Daniel E. Dombrowski, PCRA)

Valparaiso is the only suburban community in the tristate Chicago region that has had air service to another metropolitan area in the past 50 years. For a relatively brief period in the early 1980s, the community had direct flights to Indianapolis. Valparaiso's scheduled airline service, however, is best remembered for its air-taxi flights to O'Hare.

For the first 20 years after Porter County Regional Airport opened in 1949, scheduled service was only sporadically available. In 1955, Chicago Airways began flights between Midway Airport and South Bend using propeller-equipped aircraft. These flights carried cargo, mail, and passengers, but made three stops before reaching Valparaiso. As a result, they were of relatively little value to passengers.

Chicago Airways appears to have lasted only three years, but a variety of factors made the resumption of air service to Valparaiso an enticing prospect. Thousands of middle- and upper-income residents moved to the Valparaiso area in the 1960s, many of them fleeing Gary and other declining steelmaking towns. Many sought to avoid the stressful drive on toll highways to O'Hare, a distance of nearly 80 miles. To fill the void in air service, Phillips Airlines launched air-taxi service from Valparaiso to O'Hare in 1969, initially with four daily round-trips. These nonstop flights reached O'Hare in 30 minutes and were operated by small propeller-powered Piper aircraft. Flights departed from a modest terminal building built of brick and consisting of little more than a small waiting room, ticket counter, and a couple of offices.

In 1971, the flights serving Valparaiso were extended to Michigan City. (See the chapter devoted to Michigan City earlier in this section for details). Valparaiso's air service reached its peak in 1979 when a second carrier, Neimeyer Aviation,

Paper-airplane logo of Phillips Airlines. (Author's collection)

began operating weekday round-trips from Indianapolis. This carrier's small propeller aircraft touched down in Kentland, IN, before reaching Valparaiso, giving the city direct service to five communities, although two—La Porte and Michigan City—were less than 30 miles away by road.

All of these services, however, disappeared soon after. Airline travelers gradually grew more accustomed to driving to O'Hare, and competition escalated when Tri-State Coach Lines expanded its bus service in the area. Flights to Indianapolis lasted only a year before being dropped in 1980, but Phillips Airlines soldiered on until 1986, ending the last of Valparaiso's scheduled operations. The airline reportedly averaged just 18 passengers per day prior to the discontinuation of service.[87]

Valparaiso eventually lost its intercity rail and bus service as well. By the early 1990s, Valparaiso had the distinction of being the only community in the tristate Chicago region to have lost all of its long-distance service on three different modes—air, bus, and rail.[88]

The old airport terminal building was razed in 1997 to make room for a larger terminal serving the needs of general aviation. Rarely is the resumption of scheduled air service to Valparaiso seriously contemplated today, in part due to the success of South Bend Regional Airport, which is about 50 miles away. This airport's private aviation role, however, remains strong and appears poised to grow. ■

Porter County Airport, home to more than 100 private aircraft, is advertised today as "Northwest Indiana's finest all-weather business and general aviation airport." The modern terminal building (right) was built in 1997, a little more than a decade after Phillips Airlines last used the much smaller 1949 facility. (Photo by Daniel E. Dombrowski, PCRA)

WINNETKA HELIPORT
1370 Willow Rd., Winnetka, IL

Winnetka was home to the only terminal in the Chicago region that primarily served scheduled helicopter flights. The Winnetka Heliport offered a fast means of reaching Midway and O'Hare airports between 1958 and the early 1970s.

This facility emerged to support a mail-carrying operation launched by Helicopter Air Service in 1949 between the Main Post Office in downtown Chicago, Midway Airport, and several suburbs.[89] The company built a small heliport, in the form of a 200-foot-diameter circle, at 1370 Willow Rd. in Winnetka (near the northwest corner of Willow Rd. and Hibbard Rd.). This facility was surrounded by undeveloped land that mitigated the noise generated by takeoffs and landings.

Renamed Chicago Helicopter Airways (CHA), the company inaugurated passenger service between Midway and O'Hare with seven-seat Sikorsky S-55 helicopters on November 12, 1956. CHA began carrying passengers to and from Meigs Field in July 1957 and, in anticipation of expanding to Winnetka, transformed the old mail heliport into a modern passenger terminal, complete with a ticket counter, small waiting room, concrete apron, and a parking lot. CHA launched passenger service to Winnetka and Gary, IN, the following year. Fares for travel between Winnetka and O'Hare were just $5 each way (about $40 in today's dollars), and trips to Midway were just $2 more. Service was soon upgraded to 12-passenger Sikorsky S-58C helicopters.

The difficulty of driving to Midway worked to the service's advantage. Although Winnetka was a mere 13 miles by air from Midway, it was an arduous 20-mile drive on arterial highways that took at least two hours. With flights whisking passengers to O'Hare's landing pad on the rooftop of Terminal 2 in just 9 minutes, and reaching Midway in just 22 minutes, flying by helicopter could easily shave more than an hour and a half from the trip to the airport.

Advertisements boasted the suburban heliport's convenience for those living in Lake Forest, Highland Park, and Northbrook as well as Winnetka—bedroom communities for jet-setting executives. CHA added two round-trips from O'Hare to Winnetka in 1958 and studied the possibility of opening another passenger heliport serving Evanston and Skokie.[90] Winnetka's four daily round-trip flights were timed to serve business travelers. The first flight departed at 7 a.m., and the last one returned at 8:07 p.m., making same day round trips to the East Coast feasible.

The front page of *Winnetka Talk* on July 17, 1958, informed residents of the start-up of Chicago Helicopter Airways' passenger service to O'Hare. Ironically, the report about this ultra-modern service was adjacent to an article about the impending dismantling of the old North Shore Line railroad route through the community. (Winnetka Public Library)

Tragically, Flight 698 from Midway to O'Hare crashed in Forest Park, IL, in July 1960, apparently due to a structural defect. Eleven passengers and both crew members were killed, resulting in the worst accident in the history of scheduled helicopter passenger service at the time. Although this doomed trip did not originate or terminate in Winnetka, the accident hurt the carrier's image and heightened concerns over the safety of helicopter travel. The virtual abandonment of Midway as a passenger airport in the early 1960s also hurt business, as did the ease of travel on newly opened expressways. Struggling to diversify its revenues, CHA began offering express-shipment service for time-sensitive cargo in 1962, but it never fully recovered due to the decline of Midway, and only intermittently served Winnetka passengers afterwards. The company formally ended Winnetka service in January 1963.

Chicago Helicopter's last service (the route from Meigs to Midway and O'Hare) ended in 1976. Hardly a trace of the Winnetka Heliport remains today. The terminal building was razed and the concrete apron was completely removed. Baseball diamonds occupy much of the site today. ■

Chicago Helicopter Airways timetable, June 1, 1960—the start of the carrier's third year in Winnetka. (Author's collection)

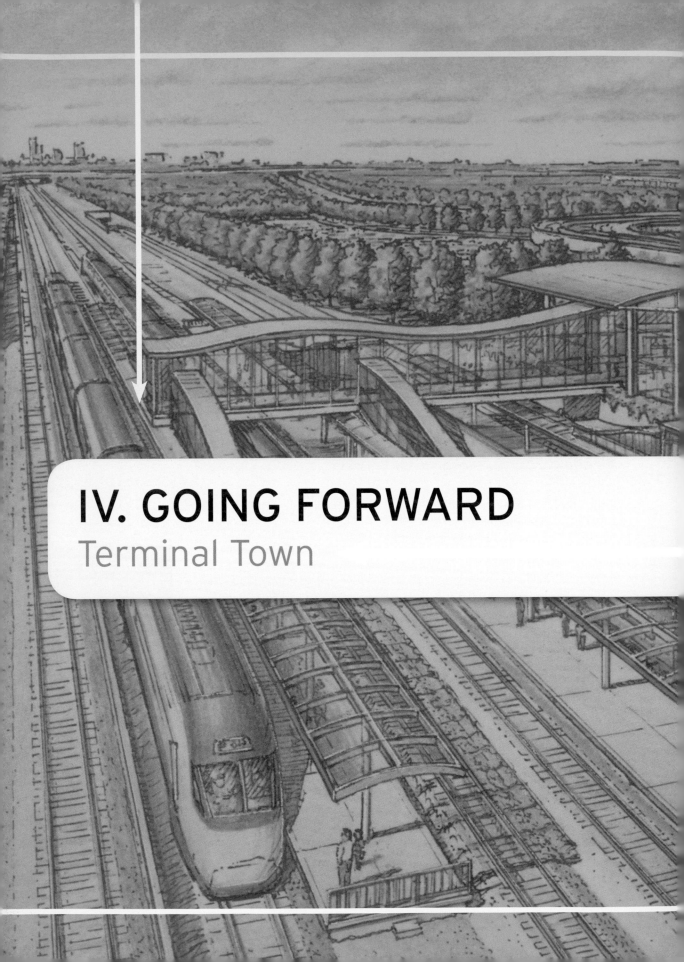

IV. GOING FORWARD
Terminal Town

THE FUTURE IN FLUX
Terminal Town

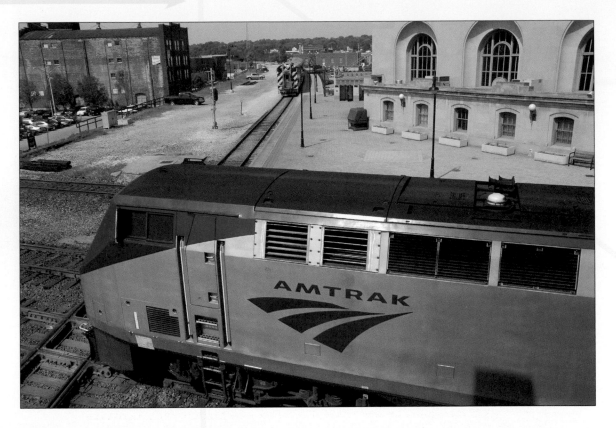

Recounting the history of the Chicago region's passenger terminals illustrates the ebb and flow of intercity transportation in one of the world's greatest travel hubs. It is not only a saga of welcoming the new and saying goodbye to the old, but also the story of how technological and economic changes have threatened the region's role as the country's preeminent transfer point.

The result has been a terminal network that has undergone a complete metamorphosis since the late 1930s. Eight airports in the metropolitan region celebrated their first scheduled passenger flights, six with helicopter service as part of the mix. Two giant new connecting complexes—O'Hare International Airport and the Chicago Greyhound Station—emerged as nationally recognized prototypes for their respective modes. Lightweight and streamlined passenger trains created excitement and optimism for long–distance train travel at the six downtown terminals.

Less than 20 years into the post-World War II era, however, it became clear that some terminals were moving closer to oblivion. The last steamer making a scheduled intercity trip from the city left from the old Goodrich Landing. Both of the main interurban railway stations adjacent to the Loop elevated were shuttered, and even the future of Midway Airport grew less certain. Then, over a nine-year period starting in 1969, five of the six downtown railroad passenger terminals ceased serving long-distance trains. Scheduled helicopter flights—heavily used in the early 1960s—were gone little more than a decade later. All of the major bus depots operating in the late 1970s were abandoned in favor of smaller stations by the mid-1990s.

This drastic change in the transportation landscape reduced some of the region's architectural treasures to rubble. Stately Grand Central Station, ambling Central Station, the streamlined National Trailways Station, Hangar One at Glenview Naval Air Station, and the modernistic Sky Harbor Airport all disappeared. Chicago's Greyhound Station; Englewood Union Station; 63rd St.–Woodlawn Station; the Winnetka Heliport; and both Midway Airport's "South Terminal" and its much larger 1947–48 replacement, while less revered by the architectural community, were all demolished with little or no objection from a largely unconcerned public.

Fortunately, significant portions of other terminals were spared from the wrecking ball. Dearborn Station's headhouse building, Glenview Naval Air Station's control tower, Navy Pier's Family Pavilion and East End buildings, North Western Terminal's suburban concourse, and Meigs Field's administration building were saved by municipally-led preservation efforts and today are widely recognized as being historically significant. Still, the region's record in protecting buildings important to its transportation heritage is hardly encouraging. While this book was being written, Aurora's Burlington Station was torn down and Gary Union Station garnered attention as one of the county's eight most endangered railroad landmarks.[92]

Among the 40 Chicago-region passenger facilities that were important terminals or transfer points in 1939, just three—Joliet Union Station, Chicago Union Station, and Millennium Station (formerly Randolph Street Station)—have been continuously used for intercity service since then.[91] The latter two, however, have been greatly modified. All of the dozens of others featured in this book were either built after World War II or lacked long-distance service for at least part of the postwar period.

Even Joliet Union Station—the only terminal that has remained both structurally unchanged and in continuous long-distance transportation use—is undergoing modifications that will enhance efficiency but greatly diminish the original structure's role as a passenger terminal. The uncharitable fate met by the Chicago region's terminals contrasts sharply with circumstances in the Northeast Corridor. There, on former Pennsylvania Railroad and New Haven Railroad lines linking Boston to the nation's capital, almost all major train stations— Boston's South Station, New York's Grand Central Terminal, Philadelphia's 30th Street Station, Baltimore's Penn Station, Washington Union Station, and others—have remained, in many respects, unchanged and in continuous service since 1940. (New York's Pennsylvania Station

(Opposite) With a Metra commuter train visible in the distance, an Amtrak train destined for Springfield, IL, and St. Louis crosses former Rock Island tracks at Joliet Union Station, circa 2010. This station is a rarity for having had no major change to its exterior and for offering continuous intercity passenger service since 1939—a combination shared by no other facility in the region. (Mark Llanuza photo)

is an exception, having been tragically demolished in 1963 but with its track arrangement largely intact, and nearby Grand Central Terminal no longer hosts intercity trains).

Historic airport terminals have also fared poorly in almost all parts of the country, although marginally better in the Northeast. Meticulously restored passenger terminals at New York's LaGuardia Airport and Ronald Reagan Washington National Airport have preserved waiting rooms and concourses that date back to the 1930s. Conversely, none of the terminal buildings at Midway built prior to 1960 survived.

As other facilities come and go, however, there can be little doubt that Chicago's "big three"—Union Station, along with Midway and O'Hare Airports—will remain important hubs for decades to come. O'Hare's terminal buildings and Union Station's Great Hall retain much of their original grandeur, making them icons of their respective modes of transportation. Both are receiving necessary investments to ready them for the needs of tomorrow's travelers.

Travelers a half-century from now will almost certainly find new stations and terminals awaiting them in the Chicago region. Much more change is on the horizon.

AIRPORTS

For the next quarter century, Midway and O'Hare are well-positioned to remain transportation stalwarts. The massive modernization programs, completed at Midway and underway at O'Hare, will allow them to absorb much of the anticipated growth in traffic in the upcoming decades. Eventually, however, the region's airport capacity will again be exhausted and the debate over airport expansion will resume.

The region's airport system will likely continue to evolve in a manner similar to that of Boston. Both cities successfully expanded their primary airports after attempts to build entirely new airports on "greenfield" sites stalled. Yet each city now faces a significant diversion of passengers to airports beyond their metropolitan regions. In Chicago, Milwaukee's Mitchell Airport—and to a much lesser extent, the Rockford and South Bend airports—offers alternative flight options. In the Boston area, airports in Hartford, CT, Manchester, NH, and Providence, RI, attract business away from Logan International Airport.

Although diversion to airports outside the Chicago region will surely continue as the population occupies an ever-widening area, there is little doubt that the push to build a third major airport will grow in intensity. The Gary/Chicago Airport and the proposed airport in Peotone, IL., remain the leading third-site contenders for the foreseeable future. Gary has the edge, but the political landscape remains far too unpredictable to anticipate the eventual outcome. Regardless, it seems highly improbable that 25 years from now air service will be as heavily concentrated at Midway and O'Hare as it is today.

The loss of Meigs Field may seem a mere footnote in this debate, but that airport's demise is noteworthy as it precludes virtually any possibility that scheduled flights by helicopters or fixed-wing aircraft will return to downtown Chicago anytime in the near future. The success of close-in secondary airports in London, England, and Toronto, ON, suggests that Meigs could have played an important strategic role in this region. As noted in Appendix I, however, an advanced heliport near Chicago's city center could partially fill the current void.

INTERCITY BUS TERMINALS

The remarkable recovery of intercity bus travel since the early 2000s makes an entirely new downtown bus terminal a distinct possibility in the not-so-distant future. Rising traffic is pushing the once-lowly intercity bus back to the forefront of passenger travel. Considerable demand exists for a bus station more centrally located than Greyhound's Chicago Express Terminal, located at 630 W. Harrison St.

When the Express Terminal opened in 1989, few seemed concerned that long-distance buses would be relegated to this peripheral location. Since then, however, bus travel has regained mainstream appeal. New city-to-city operators, particularly Megabus, are growing at a

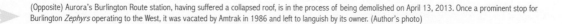

(Opposite) Aurora's Burlington Route station, having suffered a collapsed roof, is in the process of being demolished on April 13, 2013. Once a prominent stop for Burlington *Zephyrs* operating to the West, it was vacated by Amtrak in 1986 and left to languish by its owner. (Author's photo)

pace reminiscent of the airline growth that occurred after the industry's deregulation in 1979. The enormous flexibility and versatility of motor-coach travel suggest that many new services, including luxurious offerings with spacious seating and beverage service, may emerge. Many of these services will likely choose to operate from locations without traditional terminals having extensive passenger amenities.

Despite all of this growth, however, there has been little impetus in Chicago to integrate intercity bus service with other transit modes, as has been done in Boston and Washington, D.C. In these cities, intercity buses arrive and depart in retrofitted parking facilities connected to major train terminals. In Chicago, no serious proposals currently exist to re-create a bus terminal in the heart of downtown.[93]

Regardless, its seems likely that more arrival and departure locations will emerge in the Greater Loop, possibly near Millennium Park or River North, as the number of tourists who travel by intercity bus grows. Moreover, carriers operating from curbside locations are showing heightened interest in moving off city streets voluntarily. Megabus, as previously noted, is exploring the move of its hub to a location away from Union Station's perimeter on account of problems with its existing pick-up and drop-off point.

MARINE PASSENGER SERVICE

Proposals for marine passenger service across Lake Michigan are likely to emerge as entrepreneurs seek ways to profitably move passengers from Chicago to Michigan locales on the opposite side of the lake. The success of Lake Express—the high-speed catamaran-type ferry introduced between Milwaukee, and Muskegon, MI, in 2004—has whetted the appetite of those who believe that a similar service from Chicago could be viable.

High-speed ferry service, however, will remain a more difficult proposition, since the drive around the south end of the lake is more than 90 miles shorter from Chicago than it is from Milwaukee. The lack of an attractive terminal site is another issue. History has shown that Navy Pier is too far removed from the Loop, and its public transit service too limited, for it to be an ideal location for scheduled boat operators. River locations near the Loop, or insets near North Pier, could emerge as terminals for across-the-lake service, but these sites are hemmed in by development, and lack low-cost parking and easy expressway access. With fuel prices expected to remain high and the necessity of lock closures on the river during winter, the resumption of scheduled service to places outside of the metropolitan region seems doubtful over the coming decade.[94]

RAILROADS

Developing a more efficient arrangement of downtown rail terminals will likely remain a priority of transportation planners over the next 25 years. With four terminals in the city— none of which allows for significant "run-through" operations—passengers will be denied an efficient hub-and-spoke system serving both intercity trips and those within the region.

Attempting to find solutions as ambitious as those pursued by major European cites with similar dilemmas, however, may only bring frustration. In Brussels, Belgium, a deep corridor was excavated through the central city, linking its main rail stations beginning in 1952. In Berlin, Germany, nearly all commuter and intercity trains now serve one massive station, which opened in 2006. In London, a 20-mile tunnel, completed in 2010, enables Eurostar trains originating in Brussels and Paris to operate at high speeds into the city's downtown and continue to the Midlands. The Crossrail Tunnel, an even more ambitious project slated to open by 2019, will dramatically improve the efficiency of rail services through central London.

Bold projects such as these appear to be beyond the reach of Chicago-area planners for at least a decade. Yet pursuing innovative and cost-effective strategies at Union Station could dramatically enhance the station's amenities, expand the use of the venerable Great Hall, and make possible "run-through" operations using revamped versions of the mail platforms underneath the Central Post Office. Over the longer term, even more ambitious efforts, including the creation of the West Loop Transportation Center, could integrate Union and Ogilvie stations into one giant complex featuring a multi-level corridor below the street. Although the cost of this project remains beyond the identifiable funds, it is entirely conceivable that a planning breakthrough will occur over the next 20 years. Please refer to Appendix I for an overview of the vision for Union Station.

Steady gains in ridership on Amtrak, intercity bus lines, and local transit services—coupled with the state's commitment to higher-speed (110 mph) train service—create a powerful incentive for action. Accordingly, a train traveler arriving in Chicago several decades from now would expect to find station facilities much more efficient and aesthetically pleasing than those arriving today.

(Opposite) The shadow of a Frontier Airlines' twin-engine Airbus plane crosses Cicero Ave. moments after taking off at Midway Airport. This plane is destined for the carrier's Denver hub. (Lawrence Okrent photo)

BUILDING ON A SOLID FOUNDATION

Chicago's enormous achievements in transportation are recognized and celebrated by a vast array of transportation experts, enthusiasts, and writers. A 1997 *Chicago Tribune* article by Jim Sulski, a writer from Columbia College, likened these achievements to a "grand slam" in baseball since they "touch all the bases." Sulski noted, with a small dose of hyperbole, that "New York may have its subways and California may have its highways, but when it comes to transportation, Chicago has it all."[95]

The late David Schulz, of Northwestern University, maintained in this same article that "no other city in North America, and maybe the world, has had the kind of transportation predominance that Chicago has—and Chicago is barely 150 years old."[96]

Regardless of how Chicago's transportation system evolves, future generations can stand on the shoulders of giants—the generations of architects, industrialists, planners, structural engineers, and municipal officials whose collective aspirations gave the city such a magnificent set of air, bus, rail, and marine terminals. ■

(Opposite) Passengers scurry through the corridor linking Chicago Union Station's Great Hall to its passenger concourse in December 2013. Although the area's architecture has remained unchanged since the golden years of intercity passenger trains, the station's master plan calls for the restoration of many of its historic functions, including retail businesses that once catered to train travelers. (Xhoana Ahmeti photo)

This 1974 photo was taken after a wrecking ball had just laid claim to Central Station, a Windy City landmark that was the second of Chicago's "Big Six" downtown stations to be demolished. The depot's train shed had already been razed, and almost the entire site would soon be available for additional park space and redevelopment, which would eventually come in the form of the massive Central Station mixed-use project. (Verne Brummel photo)

TERMINALS OF TOMORROW

Eight Chicago-region terminal facilities envisioned by transportation planners stand out for their potential to significantly change patterns of intercity travel. At present, only a few of these projects seem imminent. Yet each has found vigorous champions and garnered the support of prominent organizations or government agencies. Many other proposals have fallen off the planning agenda. Most prominent among them are the Lake Calumet Airport on the far southeast side of the city, and the reuse of the massive Main Post Office building at 433 W. Van Buren for high-speed rail.

IMPROVING AN "ICON OF A GREAT AGE": UNION STATION

The city's only terminal to be honored as a "Great Public Place" by the American Planning Association, Chicago Union Station holds iconic status among train travelers. Although the station's track configuration and floor plan pose challenges, a transformation is underway that will greatly enhance its role as a gateway to the city. At the center of the redesign is the Great Hall, an ornate Beaux-Arts space (once the main waiting room) that has been underutilized since passenger activities were consolidated into the concourse area in 1991. This renovation improved the terminal's appearance and passenger flow, but reduced the intensity in which travelers used the Great Hall.

A wide range of strategies is helping ready the station for ever-rising passenger numbers. Amtrak, the station's owner, has made the Great Hall a focal point in its effort to rejuvenate the Headhouse Building, a six-story landmark that constitutes the western half of the station. After restoring air conditioning to the hall, which had been absent since the 1960s, Amtrak moved its Midwest regional office and operations control center to the building's restored office space in 2012. The carrier is also planning to relocate the "Metropolitan Lounge" for sleeping-car customers to the Great Hall, which, in turn, will allow for expansion of the coach-class waiting

This rendering from the Union Station Master Plan shows a possible transformation of the station's "steam tunnel" space into a passenger passageway with moving walkways and colorful art. This passageway space would connect Amtrak's waiting lounge space within the station concourse to a new waiting lounge under new passenger platforms created from the station's former mail platforms. (CDOT/Ross Barney Architects)

lounge from 450 seats to 950 seats. These enhancements could be a springboard for long-sought commercial and retail improvements in the hall.

The flow of automobile and bus traffic around the station is also being reengineered. In 2015, the Union Station Transit Center is expected to open just south of Jackson Boulevard between Canal and Clinton. This off-street facility, a project of the Chicago Department of Transportation (CDOT), will serve CTA bus routes and offer users an underground passageway to the station. CDOT will also create new traffic channels in front of the Canal entrance to separate buses from general traffic to ease curbside congestion.

Revamping the station's congested south concourse, where track and platform space are in short supply, is another priority. The Union Station Master Plan Study, launched in 2010 by CDOT in close collaboration with Amtrak and Metra, carefully examined the possibility of reusing the long-vacant old Central Post Office building immediately south of the station for passenger rail service. Concluding that there was simply no practical and desirable means for using this massive building for this purpose, it found instead that "repurposing" the old mail platforms underneath the old Post Office could boost both the station's capacity and efficiency. These disused platforms could be linked to the main passenger concourse through existing rooms and passageways that lie below.

The master plan report released in 2012 envisions making improvements in phases. Within 10 years, the plan identifies widening commuter platforms, repurposing the mail platforms for intercity passenger trains, improving street access, and reorganizing the station's concourse facilities. Computer simulation work performed for CDOT suggests these relatively modest investments could accommodate several decades' worth of growth. Over a horizon of 20 or more years, the plan calls for a large-scale expansion or replacement of station facilities, possibly even creating a new passenger waiting room above the south concourse's tracks and a multi-level subway tunnel under Clinton or Canal St. Many of these longer term ideas were drawn from the West Loop Transportation Center plan described on the following page. Regardless of how quickly these ambitious efforts proceed, the station—which the Chicago Architecture Foundation calls an "Icon of a Great Age"—is poised to remain the nerve center for Midwestern train travel.

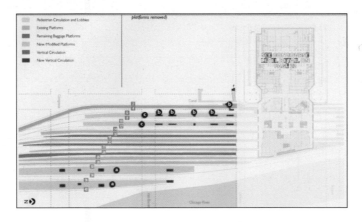

WEST LOOP TRANSPORTATION CENTER

This multi-level West Loop Transportation Center, first proposed by the Chicago Department of Transportation in 2003, would completely transform train operations at Union Station and Ogilvie Transportation Center (formerly North Western Terminal). Built entirely underground beneath either Canal St. or Clinton St., the center would have separate levels for buses, rapid-transit lines, and trains. Union Station's Great Hall would be retrofitted to serve as the main waiting room. By consolidating services of several modes, travelers could make transfers more efficiently. Trains that now terminate at either Ogilvie or Union Stations would be able to run through Chicago in order to lower operating costs and to offer more "single-seat" rides. A rapid-transit line, possibly a new CTA Red Line branch, could provide better connection opportunities for commuters than is currently possible with existing transit lines. Now part of the Master Plan for Union Station, building the center is seen as a long-term strategy likely to be pursued only after more incremental strategies have been undertaken.

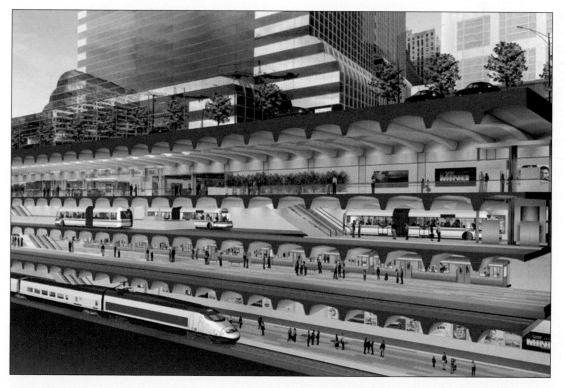

(CDOT/Transystems)

(Opposite) The Master Plan identifies removing the seldom-used baggage platforms adjacent to Tracks 8 and 12 and realigning the tracks to allow the adjacent passenger platforms to be widened (as noted by the red c's). Escalators and elevators would directly connect these platforms with street level (as noted by the red b's). Modifications to the station's mail platforms between tracks 30 and 40 (as noted by the red a's) would greatly improve the station's ability to handle long-distance traffic. (CDOT/TranSystems)

BLOCK 37 SUPERSTATION

This unfinished underground "superstation," now sealed off from the public, is designed to be the main access point in downtown Chicago for express rapid-transit service to both Midway and O'Hare Airports. Built in conjunction with the Block 37 retail-office-hotel development in 2009, the station was an integral component of Mayor Richard M. Daley's plan for fast rail service to these airports. This service, envisioned as reaching O'Hare Airport in less than 30 minutes and Midway Airport in less than 20 minutes, was seen as a potentially attractive public-private partnership. Part of a two-track connector between the CTA Red and Blue Line subway tunnels, the unfinished station follows a diagonal path northwest from the corner of State and Washington streets. Built at a cost to the city of $218 million until work stopped in 2008, it is notable for its spacious dimensions—472 feet long and 68 feet wide—and for having rows of pillars supporting a ceiling averaging 28 feet in height. This facility appears destined to remain unused until funding is identified to complete express track improvements along CTA's Blue and Orange Lines.

(Steve Serio photo, reprinted with permission, *Crain's Chicago Business.* © Crain Communications, Inc.)

O'HARE INTERMODAL TRANSPORTATION FACILITY

This massive facility is designed to foster better connections between O'Hare's auto rental services, buses, flights, and trains, as well as to lessen congestion on the airport's terminal roadways. Slated to open in 2016, the facility, formerly called the Consolidated Rental Car Facility, is located on what was formerly parking lot F. The airport's people mover is being extended by approximately 2,000 feet to provide direct service to the airport's terminals. A series of diagonal bus bays will handle coaches previously using the Bus-Shuttle Center. Although designed principally for highway-oriented modes, passengers will have only a short walk (along a landscaped path beside the consolidated rental car garage) to access Metra's O'Hare Transfer Station. This station has the potential to be upgraded to serve as a terminus for some intercity trains that presently end their run at Union Station and could one day also be a stop for Chicago–Milwaukee trains along an alternate route from that taken by today's Amtrak trains. Rail advocates see such a center as a possible incremental step to creating a bona fide high-speed rail terminal at O'Hare.

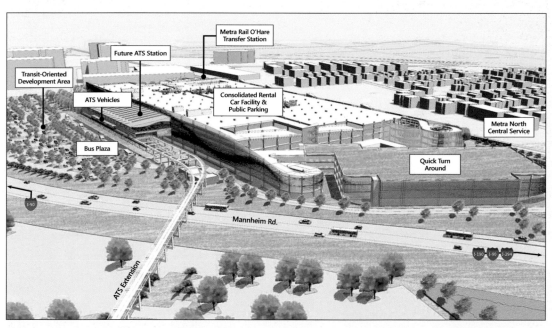

(City of Chicago/Transystems)

HIGH-SPEED RAIL TERMINAL AT O'HARE AIRPORT

A desire to integrate local air services with the proposed regional high-speed rail network has been the impetus for several ambitious proposals involving O'Hare. A recent proposal, championed by the Midwest High Speed Rail Association, focuses on high-speed trains linking O'Hare to Chicago Union Station, McCormick Place, and beyond to Champaign, IL, Indianapolis, and St. Louis. Proponents envision trains eventually operating directly into one of O'Hare's main terminals, a convenience that would likely require tunneling underneath the runways to create a track link with Metra's Milwaukee West line. This arrangement would be similar to the high-speed rail stations at Paris's Charles de Gaulle and Frankfurt's Rhein-Main airports. Alternatively, as shown on the image below, trains could operate to a dedicated terminal on the airport's east side.

(Midwest High Speed Rail Association, image by Malcomb Cunningham)

O'HARE AIRPORT WESTERN TERMINAL

A new airline terminal on the western side of O'Hare International Airport has long been envisioned as a way to expand the airfield's gate capacity and improve accessibility to suburbs located to its west. Motorists and bus travelers would be able to reach the terminal via an extension of the Elgin-O'Hare Expressway along Thorndale Rd. A new "western bypass" tollway route, looping around the western side of O'Hare, would simultaneously foster hotel and office-space development nearby. Passengers would shuttle between the terminal and O'Hare's existing terminals using an underground people mover. Some advocates even envision CTA Blue Line trains one day operating through this facility. At present, however, the O'Hare Modernization Program is moving forward without a clear timetable for the terminal, in part due to sluggish demand for new airport gates, making its construction, at best, at least a decade away.

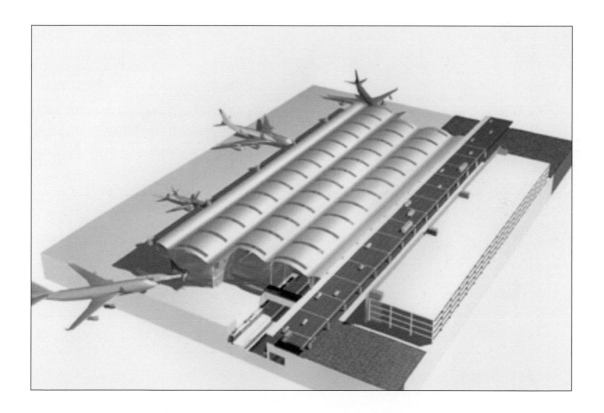

SOUTH SUBURBAN AIRPORT

Proposals for the South Suburban Airport at Peotone, IL, a small community 35 miles south of Chicago, have moved in unpredictable directions since the first technical analysis to support its construction was completed in the early 1990s. Originally conceived to resemble the massive Dulles International Airport in northern Virginia, this airport today is contemplated in less grandiose terms, in part due to the success of Midway Airport, but also due to the O'Hare expansion. The effort to build this airport has found continuous support from the State of Illinois, which has already bought a significant amount of the land needed for its construction. For several years, a conflict between Will County and a coalition led by U.S. Representative Jesse Jackson, Jr. over issues of who would control the airport inhibited prospects of federal funding. That issue was largely eliminated when the state government assumed control over the planning process in 2013. Some of the most recent proposals call for a small single-runway "starter airport" (called the Abraham Lincoln International Airport in some plans) with a four-gate terminal.

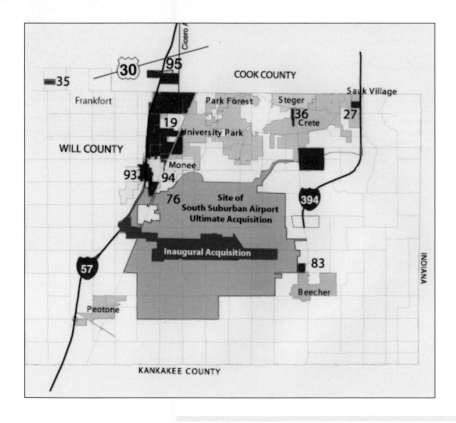

VERTIPORT ON S. HALSTED ST.

Corporations seeking to reach Chicago by helicopter complain that they have had nowhere to land in the downtown district since the demise of Meigs Field. Some of these travelers arrive at suburban airports and use limousines to finish their trips, but this oftentimes subjects them to road congestion. The push for a viable heliport downtown has grown appreciably as new pressurized "tilt-rotor" aircraft, such as the Bell 609, enter service. These aircraft are capable of traveling longer distances and operating at lower costs than conventional helicopters. A proposal for a vertiport (a heliport capable of handling tilt-rotors) at 18th St. and Damen Ave. in the Illinois Medical District stalled. In 2014, however, Chicago Helicopter Express gained city approval for a 14-helipad facility with a 17,500-square-foot hangar, a terminal, and an aircraft fueling station. Located on S. Halsted St. along the south branch of the Chicago River, the vertiport would be positioned to serve the district's hospitals as well as sightseeing and business traffic, with the possibility of service eventually scheduled to nearby and regional airports also a possibility. ■

MINOR BUS ENDPOINTS AND
AIR-TAXI STOPS

MINOR BUS AND TRAIN ENDPOINTS

In a region of such enormous complexity, it is inevitable that numerous minor terminals would emerge for intercity trains and buses. In addition to those places featured in this book, the following locations have been points of origin or final destinations for intercity buses and trains since 1939. Nonetheless, each was (or continues to be) of only minor significance to intercity travel.

Chicago. In addition to the larger facilities featured earlier in this book, at least five other locations have been points of origins and termini for regularly scheduled intercity bus and trains since 1939:

- *American Bus Lines Station* at 514 S. Wabash Ave. was the terminus for numerous intercity bus routes the carrier served, including trips from the East Coast and St. Louis. This station closed when operations were transferred to the Chicago Greyhound Station in March 1953.

- *El Expreso Bus Lines Station* at 3501 S. California Ave. has been a terminus for a route from Houston, Texas, oriented toward the Spanish-speaking population, since around 2008.

- *Jefferson Park CTA (Blue Line) stop* at 4917 N. Milwaukee Ave. was the terminus for a weekend-only Greyhound route to Urbana–Champaign, IL, from approximately 1970 to 1982. Some bus passengers may have made connections to Chicago & North Western's intercity trains stopping here through early 1971.

- *Midway International Airport* has been the terminus of Coach USA buses from northwest Indiana since the late 1980s and River Valley Transit System buses from Kankakee since 2013. Peoria Coach Bus Co. also operates to the airfield.

- *Roosevelt "L" stop* (later rebuilt as the Green and Orange Line Station) at Roosevelt Rd. and Wabash Ave. was the terminus for Chicago, North Shore & Milwaukee trains between 1938 and 1963.

Gary, IN. The Wabash Railway Station at 9th Ave. and Broadway St. served as the terminus for a mixed train (incorporating both passenger and freight cars) from Montpelier, OH, between 1931 and approximately 1959.

Landers, IL. The Wabash Railway Yard was the terminus of a mixed train from Forest, IL (on the company's St. Louis line) through the 1940s and into the 1950s.

Northbrook, IL. Northbrook Court Mall at 2171 Northbrook Ct. served as the terminus of several Greyhound routes from college campuses in Madison, Urbana–Champaign, and other points between roughly 1970 and 1980. Since the early 2000s, it has been the occasional terminus of Suburban Express/Illini Shuttle routes from Urbana–Champaign.

University Park, IL. This south suburb's Metra station has for several years been the endpoint of buses from Kankakee, IL operated by River Valley Transit System.

Several other locations, including Oak Brook Mall, appear to have been termini for certain campus bus services. The CTA (Red Line) station at 95th St., while never a terminus, has served as a stop and transfer point for Greyhound, Indian Trails, and other bus lines since the 1960s.

(Opposite) A Greyhound bus bound for Seattle is under the canopy at Jefferson Park, circa 1980. The busy CTA station served as a terminus for weekend buses operated by the carrier from Urbana–Champaign. (Mel Bernero photo)

INTERMEDIATE STOPS ON SCHEDULED AIR-TAXI ROUTES

In addition to the airports serving the Chicago region that have been originating or terminating points for scheduled airline services since 1939 (all of which are featured in Section 4), a variety of other facilities have been *intermediate stops* on air taxi flights since then. The short-lived Chicago Airways, which appears in the *Official Airline Guide* from 1955 to 1958, was the only carrier serving some of these airports. Chicago Helicopter Airways also operated a extensive system in the 1950s and 1960s. Lansing Airport or Palwaukee (Chicago Executive) Airport have for many years been prominent reliever airports specializing in general aviation but have not been served by regularly scheduled passenger airlines or air taxi providers. ■

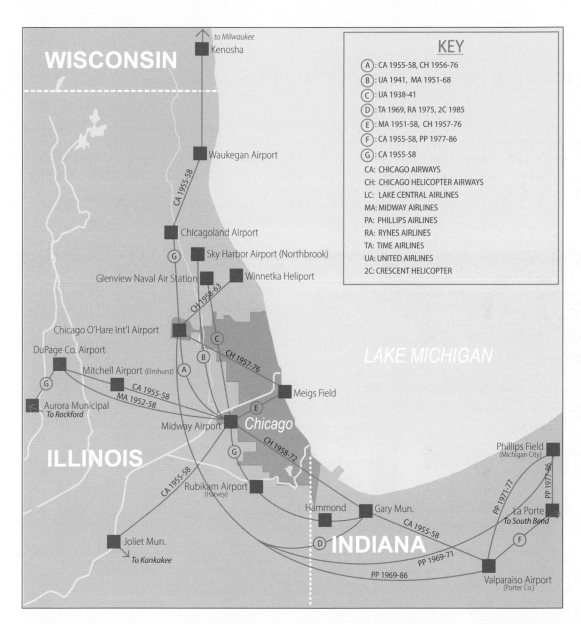

KEY

Ⓐ : CA 1955-58, CH 1956-76
Ⓑ : UA 1941, MA 1951-68
Ⓒ : UA 1938-41
Ⓓ : TA 1969, RA 1975, 2C 1985
Ⓔ : MA 1951-58, CH 1957-76
Ⓕ : CA 1955-58, PP 1977-86
Ⓖ : CA 1955-58

CA: CHICAGO AIRWAYS
CH: CHICAGO HELICOPTER AIRWAYS
LC: LAKE CENTRAL AIRLINES
MA: MIDWAY AIRLINES
PA: PHILLIPS AIRLINES
RA: RYNES AIRLINES
TA: TIME AIRLINES
UA: UNITED AIRLINES
2C: CRESCENT HELICOPTER

to Milwaukee
Kenosha

WISCONSIN

Waukegan Airport

CA 1955-58

Chicagoland Airport

Ⓖ Sky Harbor Airport (Northbrook)

Glenview Naval Air Station   Winnetka Heliport

CH 1958-63

Chicago O'Hare Int'l Airport

Ⓒ

DuPage Co. Airport

Ⓑ CH 1957-76

Mitchell Airport (Elmhurst) Ⓐ

Ⓖ

CA 1955-58

MA 1952-58

Aurora Municipal
To Rockford

Ⓔ   Meigs Field

Midway Airport   *Chicago*

ILLINOIS

Ⓖ   CH 1958-72

CA 1955-58

Rubikam Airport
(Harvey)

Hammond   Gary Mun.

LAKE MICHIGAN

Phillips Field
(Michigan City)

PP 1971-77   PP 1977-86

La Porte
To South Bend

CA 1955-58

Joliet Mun.

To Kankakee

Ⓓ **INDIANA**

Ⓕ

PP 1969-71

PP 1969-86   Valparaiso Airport
(Porter Co.)

This map shows the many locations in the metropolitan region from which this air-taxi operator offered scheduled service to Midway and O'Hare airports. The system reached its zenith in the mid-1950s, when more than a dozen locations had service, although flights to Northwest Indiana did not peak until several decades later. It is highly likely that some stops were occasionally skipped due to a lack of passengers. (Author's collection)

(Opposite) Old airplanes parked near the control tower in 2011 at Aurora's municipal airport harken back to the days when this industrial satellite of Chicago had scheduled air-taxi service to Midway. Aurora was one of eight communities in the present-day metropolitan region with short-hop flights to Midway in 1955. (Jerry Williams © 2011)

MEASURING INTERCITY SERVICE

The changing level and geographic range of transportation services from Chicago's terminals can be accurately measured by drawing upon primary sources of information. Computations of the amount of daily train activity from stations featured in this book were made using public timetables in cooperation with the Northwestern University Transportation Library and William Vandervoort, creator of the Chicago Transit and Railfan Website (www.chicagorailfan.com). Mr. Vandervoort's website is a widely recognized source of information on the region's transportation history. Other information was obtained from the Official Guide of the Railways and company timetables. Estimates of the number of bus and plane departures were made by consulting Russell's Official Motor Coach Guide and the Official Airline Guide, respectively.

All stations and terminals included in this volume met either of two criteria: they served as the origin or final destination of intercity buses, planes or trains between 1939 and the present, or they were significant points for transfers by passengers on intercity trips. Dozens of other places are (or were) connecting points involving local bus, rapid-transit, and commuter trains. These places are not included in this volume.

Readers interested in the process of estimating the extent of the services available should consult the author's white paper "Calculations of Direct Air, Bus and Rail Service from Chicago to Major U.S. Cities, 1942–present," available from the Chaddick Institute for Metropolitan Development, DePaul University (las.depaul.edu/chaddick) and from the DePaul University Library. When identifying the 100 largest cities in the country, several cities with contiguous boundaries were consolidated in a manner described in the paper.

Suggested emendations and corrections to the material in this book should be directed to chaddick@depaul.edu ■

(Opposite) Electronic departure boards show the wide range of destinations served by American Airlines at O'Hare. In the distance, a jet with the carrier's familiar unpainted polished metal fuselage provides a striking backdrop to travelers in the relatively dark interior (sfPhotocraft photo)

NOTES

[1] These 20 places consisted of the six downtown railroad stations, the Greyhound and Trailways stations, Gary Airport, Meigs Field, Midway Airport, O'Hare Airport, the three downtown interurban stations, and four outlying termini in which trains or buses from outlying points terminated (Aurora Burlington Route Station, Aurora Traction Terminal, Forest Park CA&E Station, Michigan City South Shore Line Station, and Kenosha North Shore Line station). Gary Airport is included due to its extensive service on Lake Central Airlines at the time. The five timesaving connecting points were Englewood Union Station, Gary Union Station, Hammond's bus stations, Joliet Union Station, and the Illinois Central station at 6327 S. Dorchester. All are featured in this volume. See also Appendix B.

[2] Chicago Real Estate Board, "Final Report of the Library, City Planning and Zoning Committee of the Chicago Real Estate Board on Zoning in Chicago," 1923 (Municipal Reference Collection, Harold Washington Library, Chicago), 2-12.

[3] For more information about train counts and the available service from Chicago to the 100 largest cities in the United States, see J. Schwieterman, "Calculations of Direct Air, Bus and Rail Service from Chicago to Major U.S. Cities, 1942-Present",available from the Chaddick Institute for Metropolitan Development, DePaul University (las. depaul.edu/chaddick).

[4] Ibid. In making these calculations about direct service, several cities that were contiguous with much larger cities were consolidated: Cambridge, MA (consolidated with Boston), Long Beach, CA (Los Angeles), and Waterbury, CT (New Haven, CT). The "Quad Cities" of Bettendorf and Davenport, IA, and Moline and Rock Island, IL, were combined.

[5] At one point, direct service to the Twin Cities was available on six railroads from five of Chicago's six stations: Central Station (Illinois Central and connecting roads), Grand Central (Soo Line and Chicago Great Western), LaSalle St. (Rock Island Lines), North Western Terminal (Chicago & North Western), and Union (Burlington Route and Milwaukee Road). The author thanks Fred Ash, railroad historian, for this insight.

[6] For a discussion of the advertisement, see Julian Watkins, *The 100 Greatest Advertisements 1852-1958: Who Wrote Them and What They Did,* 2012 (Dover Publications, New York).

[7] North Western Terminal and Central Station, which were separated by a mile and a half, were the farthest apart of the six major downtown railroad stations.

[8] See Robert Howard, "Reveal Talks for 13-Railroad 12 St. Station". *Chicago Daily Tribune*, March 7, 1949, for a summary of the postwar consolidation initiative.

[9] In 1957, travelers could reach 44 of the 100 largest cities directly from Union Station, with cities measured on the basis of their 1940 population. See Schwieterman", Calculations of Direct Air, Bus and Rail Service."

[10] In Chicago, only LaSalle Street Station's trains brought passengers close to the Loop Elevated system, but even this station (built by the New York Central) was on the southern periphery of the business district and had little dense development to its south or west. None of Chicago's stations were within the Loop proper; all but LaSalle were at least three blocks away.

[11] William R. Minor, "What's Wrong with Chicago's Terminals?", *Railway Progress*, July 1948, 18-21.

[12] These calculations exclude the South Shore Line, which, while offering intercity service as far east as South Bend, IN, was an interurban railway primarily serving a regional role.

[13] On the basis of the number of trains, the combined Central/Roosevelt Rd. stations held the distinction as the busiest terminal outside the Northeast U.S. even when the South Shore Line, an interurban, is exclided from its train count. In addition to the stations 58 intercity trains, it had an estimated 306 Illinois Central trains terminating in the city and Blue Island, 83 on the route to University Park, and 34 South Shore Line trains.

[14] Schwieterman, "Calculations of Direct Air, Bus, and Rail Service".

[15] The Miami-bound *Royal Palm*, which was a Big Four train out of Central Station, competed with the *City of Miami*, which was operated by the IC out of the same depot, but this train had no through coaches, requiring a transfer in Cincinnati.

[16] The Soo Line had previously used the station, between 1899 and 1912.

[17] Schwieterman, "Calculations of Direct Air, Bus, and Rail Service."

[18] Ibid.

[19] The Milwaukee Road, experiencing strong traffic growth after the demise of Chicago, North Shore & Milwaukee interurban in 1963, expanded its schedule from 54 to 71 weekday trains between 1956 and 1969.

[20] "New Union Bus Terminal to be Built in Loop: Many Lines to Use Stations,"*Chicago Daily News*, June 5, 1935.

[21] Ibid.

[22] See especially *Chicago Daily News*'s multi-page feature (Section 3 pg. 37) published on March 18, 1953. The phrase"Little City of Shops" appears in a headline on page 42. See also "Big Greyhound Bus Terminal in Loop Opened: 10 Million Dollar Unit Is Dedicated," *Chicago Daily Tribune*, March 20, 1953. Other information for this chapter was also obtained from "Chicago Bus Station," *Bus & Coach*, October 1953.

[23] "Big Greyhound Bus Terminal in Loop," 14.

[24] Paul Gapp, "End Of the Line: New Greyhound Station Suffers From Isolation," *Chicago Tribune*, March 25, 1990, 14.

[25] Robert Davis, "Leaving the Driving to Them at New Terminal," *Chicago Tribune*, December, 7 1989, 1.

[26] Paul Gapp, "End of the Line," 14.

[27] See Joseph P. Schwieterman, Lauren Fischer, Sara Smith, and Christine Towles, "The Return of the Intercity Bus: The Decline and Recovery of Scheduled Service to American Cities, 1960-2007," Chaddick Institute for Metropolitan Development Policy Study, DePaul University, December 24, 2007.

[28] The terminal was also an important arrival point for passengers from rural towns in north-central Illinois. Many made bus-to-train connections at Aurora on their way to Wells St. For additional discussion of this, see the chapter devoted to Aurora's Terminal Building.

[29] Several other wharves had scheduled departures after World War II, but these operations were less oriented toward practical transportation than the Goodrich and Navy Pier facilities and thus are not considered in this book.

[30] George W. Hilton, *Lake Michigan Passenger Steamers*," (Stanford, CA: Stanford University Press, 2002).

[31] Ibid, 158. See also *Chicago Daily News*, April 16, 1923.

[32] For a summary of postwar steamship operations, see *Lake Michigan Passenger Steamers*, 160-169.

[33] Brennan Caughron, "Five Mainline Stations on 63rd Street," *First and Fastest*, Spring 2010, 10.

[34] As noted in the section on outlying terminals, the O'Hare Bus-Shuttle Center also offers connections to passengers traveling through Chicago. Few passengers, however, make connections between buses at the center. Most arrive either to board flights or to reach destinations within the metropolitan region.

[35] Schwieterman, "Calculations of Air, Bus, and Rail Direct Service."

[36] Ibid.

[37] For example, passengers traveling between Kalamazoo and Niles, Mich., and points east of Chicago such as Cleveland, Pittsburgh, and Washington, D.C., who previously might have made connections between trains in downtown Chicago, now had more reason to transfer in Gary.

[38] "Greyhound Replies to Bus Depot Suit," *Chicago Tribune*, June 10, 1965, A3.

[39] There is a dearth of historical documentation or photographic material on Hammond's bus stations. Neither local historical organizations or motor coach advocacy groups have photos in their archives to depicting their appearance and layout. The author welcomes any information that readers can provide.

[40] Trains last operated from Joliet over Michigan Central's "Joliet Cutoff" to Lake Station, IN, in the 1920s.

[41] Michigan Central trains previously operating into Central Station were rerouted to LaSalle Street Station in 1957 and thus ceased making stops at 63rd St.

[42] The South Shore Line operates several late-night trains from South Bend that terminate in Michigan City, which might nominally fit the definition of "intercity trains." These trains, however, are operated for purposes of positioning equipment for the morning commute.

[43] The address of the bus depot was listed as both 56 N. Broadway and 54 N. Broadway in various issues of *Russell's Official Motor Coach Guide* over the years.

[44] This count includes 20 daily buses on the Joliet-Aurora Transit Lines, Inc., a privately operated local transit service.

[45] The branch, built by a CA&E subsidiary, remained in service only through 1951.

[46] The Chicago Great Western and Soo Line operated only a few trains in each direction daily and discontinued passenger service in 1956 and 1968, respectively.

[47] Forest Park was one of only three communities in the Chicago metropolitan region (Gary and Valparaiso being

the other two) that had *three* rail stations serving intercity passengers in the postwar era, but which have no intercity service today.

48 "Bulletin Board," *Chicago Tribune*, December 1, 1953.

49 "Greyhound Asks for New Evanston Bus Terminal," *Chicago Daily Tribune*, June 12, 1952, 14.

50 Anne Burris, "Travel agent wants to set up bus station in village." *Daily Herald*, May 18, 1983.

51 "Fare Wars: Bus Service Run by Student Butts Heads with Greyhound," *Daily Herald*, October 20, 1995.

52 Chicago's Midway and O'Hare airports have substantially more *originating* trips than New York City's LaGuardia and John F. Kennedy International airports. Although more traffic moves through airports in the metropolitan New York and Newark region, more trips originate from airports within the city limits of Chicago and in the New York proper. Atlanta's Hartsfield International Airport, however, generates more passenger enplanement than the airports in Chicago or New York due to the extensive number of passengers making connections through that facility.

53 Part of this problem stems from the institutional arrangements governing planning in the metropolitan region. Northwestern Indiana (which includes Gary and Hammond) and northeastern Illinois (Chicago) have separate Metropolitan Planning Organizations as well as different organizations to plan and fund transit operations.

54 For a discussion of the political aspects of airport development in the region, see Joseph Schwieterman and Alan Mammoser, *Beyond Burnham: A History of Planning for the Chicago Region*," (Lake Forest, IL: Lake Forest College Press, 2009). ·

55 Ibid.

56 Midway claimed to be the "World's Busiest Airport." New York Municipal (LaGuardia), however, had more flights listed in the *Official Airline Guide* between the World War II years through 1947. After that year, and continuing through 1962, Chicago Municipal (Midway) had the most flights.

57 For an illustrated history of Municipal Airport's early years, see Chris Lynch, *Chicago's Midway Airport: The First Seventy Five Years* (Chicago: Lake Claremont Press, 2002).

58 This estimate is based on schedule information published in "Consolidated Air Travel Schedules & Fares", issued by the American Air Transport Association, Fall 1929.

59 Schwieterman, "Calculations of Direct Air, Bus, and Rail Service."

60 See Wayne Tomis, "Fight over Midway's Restoration Looming," *Chicago Daily Tribune*, February 11, 1963.

61 Midway saw the restoration of international service in the 1980s with Canadian Airlines jet flights to Toronto, and more recently with Toronto-based Porter Airlines' turbo-prop service. International flights also serve Guadalajara, Mexico.

62 See Fran Spielman, "Midway Airport's privatization prospects still unclear," *Chicago Sun-Times*, July 13, 2012.

63 *Chicago Daily Tribune*, October 31, 1955, 1.

64 See "Carson's Martin Heads for 8 Week Europe Tour," *Chicago Daily Tribune*, September 15, 1959. William W. Yates, "Direct Jet Service to Slice Chicago-to-Europe Time," *Chicago Daily Tribune*, February 21, 1960.

65 Schwieterman, "Calculations of Direct Air, Bus, and Rail Service."

66 Examples of cities that received new nonstop service include Charleston, W. VA, Colorado Springs, CO, Lexington, KY, Little Rock, AR, and Reno, NV.

67 These flights made intermediate stops in Lafayette and Kokomo, Ind.

68 Rynes Airlines began with seven flights a day between O'Hare and Gary using Twin Beech aircraft on July 28, 1975. These flights operated for only a few months before being discontinued. They had disappeared from the Official Airline Guide *by July 1976.*

69 A short-lived helicopter service operated by Crescent Helicopters appears to have operated to O'Hare in the mid-1980s. Schedules for Crescent appeared in the *Official Airline Guides* in January 1985 with three weekday round-trips to O'Hare.

70 Flights were provided by twin-engined Hawker Siddeley HS-748 turbo-prop aircraft with 48 seats.

71 Chicago Helicopter Airways suspended service on several occasions before permanently discontinuing it in 1976.

72 Muskegon service was provided as part of a route to Grand Rapids.

73 This Amtrak station opened in 2005 and is connected to the airport's main terminal via a shuttle bus.

[74] This service was initially provided by Peoria-Rockford Bus Co. The present operator is Van Galder, Inc., a subsidiary of Coach USA.

[75] "Air Taxis to Link North Shore and Loop to Airport: Service Will Connect with Regular Planes," *Chicago Daily Tribune*, March 24, 1938.

[76] Advertisements for Chicago Airways first appeared in the *Official Airline Guide* in 1955. No addition information could be found to ascertain the dates this carrier started and stopped service.

[77] The closing date was reported by Mike Schelter on the web site "Abandoned the Little Used Airfields," accessible at www.airfields-freeman.com/IL/Airfields_IL_Chicago_NW.html#chicagoland.

[78] "Okay Daily Flights from Chicago to DuPage Airport," *Bensenville Register*, June 30, 1952, 1.

[79] "New Helicopter Link Planned for Three Airports," *Chicago Tribune*, November 1, 1984.

[80] "United Launches Air Connection for East from N.S.," *Chicago Daily News*, February 15, 1938. There is no record of these flights in the *Official Airline Guide*, although that is likely attributable to the fact that air-taxi services were not listed in this publication prior to World War II.

[81] The Glenview service is listed in United Air Lines timetables of this era but asks customers to "call the airport agent" for flight times. United's April 27, 1941, timetable indicates that this service had been moved to Sky Harbor Airport.

[82] This claim was made by Elizabeth J. Murray, "Flashbacks", *Flying Magazine*, October 2001, 140.

[83] The Midway frequencies are based on scheduled operations on Fridays. All but one of these flights operated at least daily except Saturday.

[84] The company maintained its headquarters on Dundee Rd. in Northbrook. See "First Air Taxis to Begin with Flights Today," *Chicago Daily Tribune*, June 29, 1951.

[85] Charles Ballenger and Richard Dunlop, "Vest Pocket: The World's Shortest Airlines Carriers Commuters to and from the Loop at 150 mph." Undated magazine article provided to the author by the Midway Historians.

[86] "Shuttle, Bulletin Board", *Chicago Daily Tribune*, September 18, 1953, C9.

[87] See "*Porter County Regional Airport-History*," posted on www.vpz.org/history.php and accessed on September 19, 2012.

[88] Throughout all or parts of the 1960s, Valparaiso had three passenger railroads—Grand Trunk Western, Nickel Plate Road/Norfolk & Western, and Pennsylvania Railroad/Penn Central—as well as Greyhound buses and scheduled air service. Penn Central, followed by Amtrak, provided service over the former Pennsylvania Railroad route. By the early 2000s, the last of the above services had been discontinued.

[89] "Mail Test Runs by Helicopters to Begin Today," *Chicago Daily Tribune*, October 1, 1946.

[90] "Study Plots Heliport for North Shore: Evanston Area May Get Unit," *Chicago Daily Tribune*, February 8, 1959.

[91] Millennium Station's only intercity route is the South Shore Line's corridor to Sound Bend, which is heavily used by commuters. Milwaukee's Mitchell Field and South Bend's Regional Airport have also had continuous service during this period. Their emergence as important options for travelers living in Chicago is a relatively recent phenomenon.

[92] "NRHS Announces Eight Top Endangered U.S. Railroad Landmarks," press release by National Railway Historical Society, April 25, 2013.

[93] The migration of business to Wacker Dr. and the reconfiguration of this two-level roadway has made the route used by buses accessing the Chicago Greyhound Station less attractive than it was years ago.

[94] The Great Lakes cruise business, heavily acclimated toward heritage tourism, is a popular way to experience cities along Lake Michigan. Due to the relatively slow speed of travel, however, it is not competitive as a means of practical transportation. For the foreseeable future it appears that scheduled marine service in Chicago will be dominated by river taxis between points in the downtown area.

[95] By Jim Sulski, "Grand Slam: No Other City Touches All the Transit Bases Like Chicago," *Chicago Tribune*, October 26, 1997.

[96] Suburban Express/Illini Shuttle and LEX Express (which no longer operates) have both operated from numerous suburban retail areas, including Oak Brook Mall and Louis Joliet Mall in Bolingbrook, IL. The timetable observed, however, do not suggest these locations were ever the origin or terminus of regularly scheduled routes. Greyhound appears only to have originated trips from Elk Grove Village, Jefferson Park, Northbrook Court, and Woodfield Mall.

BIBLIOGRAPHY

AIRPORTS

Branigan, Michael. *A History of Chicago's O'Hare Airport*. Charleston: The History Press, 2012.

Brodherson, David. "'All Airplanes Lead to Chicago': Airport Planning and Design in a Midwest Metropolis." *Chicago Architecture* 1872-1922 Birth of a Metropolis, edited by John Zukowsky,75-97. Chicago: The Art Institute of Chicago, 1993.

Davies, Ronald E.G., and Imre E. Quastler. *Commuter Airlines of the United States.* Washington, D.C.: Smithsonian Institution Scholarly Press, 1995.

Kent, David E. *Midway Airport (Images of America)*. Mount Pleasant, S.C.: Arcadia Publishing, 2013.

Lynch, Christopher. *Midway Airport: The First 75 Years.* Chicago: Lake Claremont Press, 2002.

Official Airline Guide, North America Edition, various issues, 1939-present.

Schwieterman, Joseph P. and Alan Mammoser. *Beyond Burnham: An Illustrated History of Planning for the Chicago Region.* Chicago: Lake Forest College Press, 2009.

Selig, Nicholas C. *Lost Airports of Chicago*. Charleston, S.C.: The History Press, 2013.

Young, David. *Chicago Aviation: An Illustrated History*. DeKalb: Northern Illinois University Press, 2003.

BUS STATIONS

Russell's Official Motor Coach Guide, Des Moines, Iowa: Russell's Guides, 1939-present

Vandervoort, William V. "Classic Chicago Bus Stations," accessed on July 5, 2013, www.chicagorailfan.com.

Walsh, Margaret. *Making Connections, The Long-Distance Bus Industry in the USA*. Aldershot, U.K.: Ashgate Publishing Limited, 2000

RAILROAD AND INTERURBAN RAILWAY STATIONS

Ash, Fred W. "A History of C&NW's Chicago Passenger Stations, Part II: Birth of Madison Street Station." *North Western Lines* (1991): 20-40.

Ash, Fred. "A History of C&NW's Chicago Passenger Stations, Part III." *North Western Lines* (1991): 30-46.

Ash, Fred W. "Chicago's Legendary Roosevelt Road: Still the Place to See 1,000 Trains a Day," *Trains 73* (2013): 24-33.

Bach, Ira. and Susan Wolfson. *A Guide to Chicago's Train Stations: Present and Past*. Athens, Ohio: Shallow Press, 1986.

Burke, Tom and Graham Garfield, "Arthur Gerber: Insull's Transit Architect." *First & Fastest: Quarterly Magazine of the Shore Line Interurban Historical Society* 23 (2007): AG1-AG27.

Burnham, Daniel H. and Edward H. Bennett. *Plan of Chicago.* Edited by Charles Moore, Introduction by Kristen Schaffer. New York: Princeton Architectural Press, 1993.

Carlson, Norman. "Wells Street Terminal." *First & Fastest* 19 (2003): 26-31.

"Chicago "L".org." accessed on August 15, 2013, www.Chicago-L.org .

Condit, Carl W. Chicago, *1910-29: Building, Planning and Urban Technology.* Chicago: University of Chicago Press, 1973.

DeRouin, Edward M. *Chicago Union Station: A Look at its History and Operations Before Amtrak.* Geneva, IL: Pixels Pub, 2003.

Grant, H. Roger. *The North Western: A History of the Chicago & North Western Railway.* DeKalb: Northern Illinois University Press, 1996.

Krambles, George and Arthur Peterson. *CTA at 45*. Oak Park: George Krambles Scholarship Fund, 1993.

Lowe, David. *Lost Chicago.* Boston: Houghton Mifflin, 1975.

Hamlin, George W. *Chicago Railroad Scenes in Color (Volume 2)*, 1970-71. Hanover, PA: The Railroad Press, 2012.

Holland, Kevin J. *Classic American Railroad Terminals.* Minneapolis: Motor Books, 2001.

Hunt, D. Bradford and John B. DeVries. *Planning Chicago.* Chicago: American Planning Association, 2013.

Keating, Ann Durkin. *Chicagoland: City and Suburbs in the Railroad Age.* Chicago: University of Chicago Press, 2005.

Kelly, John. *Chicago Stations & Trains Photo Archive.* Hudson, WI: Iconografix, 2008.

Lowe, David. *Lost Chicago.* Boston: Houghton Mifflin, 1975.

Official Guide of the Railways and Steam Navigation Lines of the United States, Puerto Rico, Canada, Mexico andCuba. Newark, N.J.: JoC, 1939 - present.

Plous, F.K. *Railroad Stations of Chicago,* unpublished manuscript, 2012.

Schwieterman, Joseph P. "Ticket to Everywhere: The Extraordinary Legacy of Chicago's Downtown Railroad Terminals." *First & Fastest* 24 (2008): 8-13.

Young, David. *Chicago Transit: An Illustrated History.* DeKalb: Northern Illinois University Press, 1998.

Vandervoort, William V. "Classic Chicago Bus Stations." Accessed on July 5, 2013. www.chicagorailfan.com.

Young, David. *The Iron Horse and the Windy City: How Railroads Shaped Chicago.* DeKalb: Northern Illinois University Press, 2005.

Zukowsky, John. *Chicago Architecture, 1872-1922: Birth of a Metropolis.* New York: Prestel Publishing, 2000.

Zukowsky, John. *Chicago Architecture and Design, 1923-1993: Reconfiguration of an American Metropolis.* New York: Prestel Publishing, 2000.

STEAMSHIP TERMINALS

Hilton, George W. *Lake Michigan Passenger Steamers.* Stanford, CA: Stanford University Press, 2002.

Moffat, Bruce G. "Streetcars at Navy Pier." *First & Fastest* 21 (2005): 34-39.

Young, David M. *Chicago Maritime: An Illustrated History.* DeKalb: Northern Illinois University Press, 2001.

ADDITIONAL SOURCES

The following references, sorted alphabetically by facility, are among the more significant company documents, newspaper and journal articles, books, and other ancillary sources used in this historical compilation.

AIRPORTS

Chicago-Rockford Airport:

Chicago Tribune, December 14, 1997; Chicago/Rockford International Airport, accessed August 1, 2012, http://www.flyrfd.com.

Chicagoland Airport:

Chicago Airways: Official Airline Guides (Advertisements), North America editions; Schelter, Mike. "Abandoned & Little Known Airfields," accessed on August 15, 2013, www.airfields−freeman.com/IL/; See especially Selig (2013), 117-120.

DuPage Airport:

Ballenger, Charles and Richard Dunlop. "Vest Pocket: The World's Shortest Airlines Carriers Commuters to and from the Loop at 150 mph." Chaddick Institute Library. (n.d.) 17-19; *Chicago Tribune*, November 1, 1984.

Gary-Chicago Airport:

Chicago Tribune, May 23, 1953; August 13, 1958; February 2, 1958; November 23, 1958; November 1, 1984; February 3, 1985; December 27, 1987; August 21, 1992; February 23, 2001; *Gary Post-Tribune,* April 21, 1992; February 23, 2001.

Glenview Naval Air Station:

BIBLIOGRAPHY

Dawson, Beverly Roberts. *Glenview Naval Air Station (Images of America)*. Charleston: Arcadia Publishing, 2007; *Chicago Tribune*, February 15, 1938; March 24, 1938; Glenview View, various issues in Glenview Historical Society Collection; United Airlines: Public Timetables, 1938-41.

Meigs Field:

Midway Historians, discussions with author, May 2013; Illinois General Assembly. 1972. *Intrastate air operations in Illinois*. Chicago: State of Illinois, 5-25; Keller, David. "Ozark Airlines Twin Otters," accessed on April 13, 2013, http://airlinetimetableblog.blogspot.com/2010/11/ozark-air-lines-twin-otters.html.

Michigan City Airport:

Philips Family members, email correspondence, June 2013.

Milwaukee-General Mitchell:

Hardie, George A. *Milwaukee County's General Mitchell International Airport - A Record of Progress*. Milwaukee: Friends of the Mitchell Gallery of Flight, 1996; Dawson, Beverly Roberts. *Glenview Naval Air Station (Images of America)*. Charleston: Arcadia Publishing, 2007; Historical display, General Mitchell International Airport, viewed Summer 2013.

Midway Airport:

Chicago Tribune, February 11, 1963; *Chicago Sun-Times*, July 13, 2012; Lynch, Christopher. *When Hollywood Landed at Chicago's Midway Airport: The Photos and Stories of Mike Rotunno*. Charleston: The History Press, 2013; Midway Airport Historians, discussions with author, January 1-August 31, 2013; See especially Kent (2013), 63-96; and Lynch (2002), 1-14, 109-149.

O'Hare Airport:

Bensenville Register, October 18, 1962; *Chicago Tribune*, October 31, 1955; September 15, 1959; See especially Brodherson (1993), 75-97; and Young (2003).

Sky Harbor Airport (Northbrook):

Ballenger, Charles and Richard Dunlop. "Vest Pocket: The World's Shortest Airlines Carriers; Commuters to and from the Loop at 150 mph." Chaddick Institute Library. (n.d.) 17-19; Mercurio, Frank. "Modern Before Mies: Sky Harbor," accessed on October 2, 2013, http://modern-b4-mies.blogspot.com/2011/12/sky-harbor-airport-of-tomorrow.html; *Chicago Tribune*, October 1, 1946; June 29, 1951; Hughes, Judith Joslyn and Karie Angell Luc. *Northbrook (Images of America)*. Charleston: Arcadia Publishing, 2008; See especially Selig (2013), 47-50.

South Bend Regional Airport:

Historical display, South Bend Regional Airport, viewed on June 4, 2013; "Glass in the Modern Airport," *Architectural Forum*, November 1949.

Valparaiso (Porter Co.) Airport:

Chicago/Rockford International Airport, accessed August 1, 2012, http://www.flyrfd.com; Photographic archival materials, Porter County Airport, viewed on October 1, 2013; Porter County Regional Airport - History, accessed on September 19, 2012, www.vpz.org/history.php.

Winnetka Heliport:

Chicago Helicopter Airways: Public Timetables; *Chicago Tribune*, October 1, 1946; September 20, 1949; February 8, 1959; May 25, 1962.

BUS STATIONS

Chicago Greyhound Station (1953):

"Chicago bus station," *Bus & Coach* 25, no. 300, (October 1953); *Chicago Tribune*, Feb. 8, 1953; March 18, 1953; March 20, 1953; March 25, 1990; "Greyhound Opens New Terminal," *Bus Transportation*, March 1953.

Chicago Greyhound Terminal (1989):

Chicago Tribune, Dec. 7, 1989; March 25, 1990.

Chicago Trailways Station:

Chicago Tribune, June 5, 1935; July 3, 1936.

Crestwood Coach USA Terminal:

Chicago Tribune, August 13, 1967; September 7, 1967.

Hammond Greyhound and Trailways Stations:

Chicago Tribune, January 10, 1965.

O'Hare Bus/Shuttle Center:

Chicago Tribune, February 24, 1996; April 10, 1996.

Union Bus Station:

Chicago Tribune, June 3, 1943; July 16, 1943; South Loop Historical Society. "Union Bus Depot Now Open," accessed on July 6, 2013, http://www.southloophistory.org/transportation/unionbusdepot.htm.

Woodfield Mall:

Chicago Tribune, September 21, 1978; Daily Herald, May 18, 1983; October 20, 1985; February 24,1996; December 14, 1997; December 3, 2003.

RAILROAD AND INTERURBAN RAILWAY STATIONS:

Adam/Wabash North Shore Line Station:

North Shore Line: Public Timetable, various issues; South Loop Historical Society. "Union Bus Depot Now Open," accessed on July 6, 2013, http://www.southloophistory.org/transportation/unionbusdepot.htm; Adams/ Wabash, accessed July 1, 2012, www.chicago-l.org/stations.

Aurora Burlington Route Station:

Spoor, Michael. "Chicago & Aurora Divisions," accessed on June 1, 2013, www.chicagoandauroradivisions.com.

Aurora Traction Terminal:

Carlson, Norman. "Five Railroads and River Named Fox," *First & Fastest* 23, no. 2 (Summer 2007).

Central Station:

Lind, Alan R. *Limiteds Along the Lakefront: The Illinois Central in Chicago.* Park Forest, IL: Transport History Press, 1986; See especially Holland (2001), 71-74.

Chicago Union Station:

Fred Ash, interview with the author, December 2011; See especially DeRouin (2003); and Holland (2001), 87-91; City of Chicago Department of Transportation, Chicago Union Station Master Plan Study, May 2012.

Dearborn Station:

Kanary, George E. "Riding the Dolton Cannonball," *First & Fastest* 25, no. 4 (Winter 2010): 14-15; Kosik, Bob. "Dearborn Station," *The Warbonnett: Official Magazine of the Santa Fe Railway Historical Society* 6, no. 2 (Summer 2000): 11-26.

Englewood Union Station:

Bernard, Marty, "A Morning at Englewood Union Station," *First & Fastest* 26, no. 1 (Spring 2010): 9, 11-15; Caughron, Brennan. "Five Mainline Stations on 63rd Street," *First & Fastest* 26, no. 1 (Spring 2010): 10; *Chicago Tribune*, August 28, 1898; Ingles, J. David. "Rocketing on By." *Classic Trains* 6, no. 3, (Fall 2005): 68.

Forest Park CA&E Station:

Weller, Peter. *The Living legacy of the Chicago Aurora and Elgin: An illustrated history of the CA&E and its transition to the Illinois Prairie Path*, San Francisco: Forum Press, 1999.

Grand Central Station:

Ash, Fred. "Chicago's Grand Central Station, Part III." *Sentinel: Quarterly Magazine of the B&O Historical Society* 16, No. 2, 2-15; *Chicago Tribune*, April 30, 1969; May 4, 1969.

Gary Metro Center:

Gary Post-Tribune, December 27, 1987; June 6, 1989; *Times of Northwest Indiana*, November 23, 2005.

Hammond Monon Station:

BIBLIOGRAPHY

Dolzall, Gary W. and Stephen F. Dolzall. Monon: *The Hoosier Line,* Bloomington: Indiana University Press, 2002; Monon Railroad Historical & Technical Society. "Hammond Monon/Erie Station, Bygone Places on the Monon," accessed on April 2, 2013, www.monon.monon.org.

Howard St.:

Chicago Tribune, June 12, 1962; Duke, Donald, "North Shore Line's Last Summer: Part 2", *First & Fastest* 23, no. 1 (Spring 2007): 14-17.

Joliet Union Station:

Blackhawk Chapter, National Railroad Historical Society. "Joliet Union Station." Accessed on April 5, 2012. www.blackhawknrhs.org; See especially Holland (2002), 80-82.

Kenosha North Shore Line Station:

Burke, Tom. "The Transit Stations of Arthur Gerber," accessed on April 12, 2013, www.glenviewcreek.com/stations; Kenosha Streetcar Society, (ed.). *Kenosha On the Go*, Charleston: Arcadia Publishing, 2008, 8-9, 66.

LaSalle Street Station:

Wojtas, Edward J. "Rock Island Memories," *Passenger Train Journal* 23, no. 2 (February 1992): 24-29; See especially Kelly (2008), 67-78; and Holland (2001), 77-82.

Michigan City South Shore Line Station:

Burke, op cit. Accessed on June 1, 2013.

North Western Terminal:

See especially Ash (1991) and Grant (1996)

Randolph Street Station:

Carlson, Norm. "Train from Randolph Street: 150 years of service on the Illinois Central and Metra Electric," *First & Fastest* 22, no. 2 (Summer 2006): IC1-IC16; Lind. op cit., 7-12, 84-87, 4-7, 99.

Wells Street Terminal:

Wells Street Terminal, accessed July 1, 2012, www.chicago-l.org/stations; "The Great Third Rail in Grand Central Station or Randolph Street Stations," *First & Fastest* 26, no. 2 (Summer 2003): 3-8; See especially Carlson (2003).

63rd-Dorchester Station:

Chicago Tribune, July 17, 1969; April 9, 1963; "Concrete in New Office Building and Station," *Engineering & Cement World* 12, no. 2 (January 1, 1918): 60.